Tomato

A Complete Guide

AGRIHORTICO CPL

Copyright © 2019 **AGRIHORTICO**
All rights reserved.
www.agrihortico.com

Table of Contents

TOMATO: AN INTRODUCTION 1

COMMERCIAL GROWING PRACTICES FOR TOMATOES 11

POST HARVEST PRACTICES FOR TOMATOES 25

ORGANIC GROWING PRACTICES FOR TOMATOES.........................27

HYDROPONIC GROWING PRACTICES FOR TOMATOES 41

GREENHOUSE GROWING PRACTICES FOR TOMATOES 51

CONTAINER GARDENING PRACTICES FOR TOMATOES 61

NUTRITION AND HEALTH BENEFITS OF TOMATOES 67

FOOD USES OF TOMATOES 73

PROCESSED TOMATO FOODS 79

Tomato: An Introduction

Scientific name of Tomato is *Lycopersicon esculentum*. Tomato is the world's largest vegetable crop after potato and sweet potato. It belongs to the Solaneceous family, along with tobacco, potato and bell pepper. Tomato is native to Peruvian and Mexican region. It is considered as a very important vegetable due to its high nutrient value. The wide versatility in its usage makes it a favorite vegetable among food consumers and food processors alike.

There are thousands of different types of cultivars and varieties of tomatoes. There are mainly *heirloom tomatoes* and *hybrid tomatoes*. What we get in the markets or the regular type tomatoes are mostly hybrid tomatoes that are carefully bred for specific edible characteristics.

Figure 1: Hybrid Tomato Plant

Heirloom tomatoes are naturally-occurring, open-pollinated non-hybrid varieties of tomatoes. Sometime these tomatoes are referred as 'heritage tomatoes'. These plants are not pest-disease resistant or high-yielding as commercial hybrid tomato cultivars. Their fruits have very less shelf life as compared to that of commercial cultivars.

Figure 2: Heirloom Tomato Plant

Tomato plants are suited for growing in a wide range of climates. It can be grown in greenhouses throughout the year. Tomato plants are suitable for hydroponic growing also. Tomato is the one of the most popular and largest cultivated vegetables with wide variations in size, colour and shape. Size varies from small (tiny tomatoes) to large (plum tomatoes) while shape varies from round, oblong and globe.

Figure 3: Different Shapes and Sizes of Tomatoes

Tomatoes are available in many colours also such as green, pink, red, yellow, and orange. There are blue, dark purple and black tomatoes as well.

Figure 4: Different Colours of Tomatoes

Tomatoes are also one of the largest canned vegetables. Tomatoes are also known for its highest nutritional value. Tomato is counted among the top 50 nutrient-dense, plant-based foods. In fact, tomato is the most popular and less expensive superfood available to us today.

Taxonomy: Detailed taxonomic classification of tomato plant is given below:

Kingdom	Plantae
Order	Solanales
Family	Solanaceae
Genus	Lycopersicon
Species	esculentum

Origin and Distribution: Tomatoes are believed to be originated in South America. Today tomato is cultivated in every nook and corner of the world, in open fields, greenhouses, commercial farms, home gardens, and in hydroponic grow systems. Major commercial producers of tomatoes are China, Europe, India, and the USA.

Plant Description: Tomato plants are grown for its fruits, i.e. tomatoes. Tomatoes are botanically fruits called 'berries' but considered as culinary vegetables as they are used as vegetables rather than as fruits.

Two types of tomato plants are widely grown. These are **Determinate** (bush type) tomato plants and **Indeterminate** (vine type) tomato plants. Bush-type plants are annuals that stop growing at a certain height presenting a bush-like appearance; their stems are strong and they produce only one crop, *all at once.*

Determinate Tomatoes or Bush Tomatoes: These are dwarf, bushy, early-maturing type of tomato plants that are easy to grow and need no pruning and training. They are adapted to cool growing environments and are suitable for container gardening.

Figure 5: A Bush Tomato Plant

Some popular varieties in this category are 'Celebrity', 'Better Bush', 'Siberian', 'Beaverlodge', 'Glacier', 'Early Girl', 'Mountain Spring', 'Sophie's Choice', 'Bushsteak' and 'Ida Gold'.

Indeterminate Tomatoes: These tomato plants grow up to 3 meters in height; stem is tender and vine-like and often needs support; these plants are perennials but can be cultivated as annuals. Fruits are available throughout the year and therefore indeterminate tomato plants are preferred for commercial cultivation. Some popular varieties in this category are Sun Gold, Big Boy, Sweet Million, and Beef Steak.

Figure 6: Indeterminate Tomato Plants

Semi-Determinate Tomatoes: There are some tomato varieties which are having both the characteristics of determinate and indeterminate tomatoes. For example, Roma and San Marzano.

Commercial Varieties: Based on the size of the fruits, tomatoes are classified into 7 groups such as beefsteak tomatoes, plum tomatoes, cherry tomatoes, grape tomatoes, Campari tomatoes, Tomberries, and globe tomatoes.

1. Beefsteak tomatoes: These are large tomatoes having 4 inches or more in diameter, and are round shaped with thin skin. Beef steak tomatoes are mainly used for sandwiches. They are indeterminate type of tomatoes.
2. Plum tomatoes: They are thick and fleshy, oblong-shaped tomatoes having 3-4 inches length and 2 inches diameter with high TSS (total soluble solids). They are suitable for canning, and for making sauce and tomato paste. They are semi-determinate type of tomatoes. For example, Roma
3. Cherry tomatoes: These are small, round tomatoes of cherry-size; they are with less than one inch in diameter and sweet in

taste. Some popular cultivars are Patio Choice Yellow, and Tumbling Tom Yellow.

Figure 7: Cherry Tomatoes

4. Grape tomatoes: These are small, round tomatoes of grape-size; they look like tiny variations of plum tomatoes
5. Campari tomatoes: These tomatoes are bigger than cherry tomatoes, but smaller than plum tomatoes.
6. Tomberries: They are very tiny tomatoes having about 5 mm in diameter.
7. Globe tomatoes: These are tomatoes of average size, globe-shaped and having a diameter between 2 and 2.5 cm. They are commercially cultivated for processing and fresh consumption purposes.

Blue Tomatoes, Black Tomatoes and Tigerella Tomatoes

Blue Tomatoes: These are artificially bred tomato cultivars with high levels of anthocyanins, a group of health-enhancing antioxidants. The most popular blue tomato cultivar is 'Indigo Rose'. Another popular cultivar from Israel is 'Black Galaxy'.

Figure 8: Blue Tomatoes

Black Tomato: It is an heirloom cultivar of tomato originating from the Crimean peninsula. The most popular black tomato cultivar is 'Black Krim' which is dark reddish-purple to black in colour with dark green or brown shoulders.

Figure 9: Black Krim Tomato

Tigerella Tomato: It is a bicoloured heirloom tomato cultivar which is red in colour with yellow stripes. It is a small cultivar of tomato with a sweet flavour.

Figure 10: Tigerella Tomato

Green Zebra: This is another bicoloured tomato cultivar. It is an artificially-bred, ready-to-eat cultivar of tomato. These tomato fruits are characterised with dark green stripes on yellow coloured fruit skin.

Figure 11: Green Tomatoes

Commercial Growing Practices for Tomatoes

For large-scale commercial cultivation, tomatoes are grown as a field crop. A detailed account of growing practices for tomato plants in the open fields is given below:

Figure 12: A Tomato Garden

Climate and Soil: Tomato is a warm season crop and is susceptible to frost injury. Therefore frost protection is necessary during winters. Tomatoes may be grown at temperatures ranging from 18 °C to 27 °C. Tomatoes favour direct sunlight. Variations in temperature and light intensity may affect the fruit-set, pigmentation and nutritive value of the fruit. The best soil for tomato cultivation is well-drained, rich, fertile loamy soils. Optimum soil pH is 6.0 to 7.0.

Sowing and Planting: Seeds are sown in well-prepared nursery beds. For winter crop in tropics, seeds may be sown in June-July and for the spring-summer crop, seed are sown in November. Two to three sowings per year can be done in regions with mild climate. In the hills of tropics, seeds are sown in March-April. There are about 300 seeds in one gram. About 400 to 500 grams of seeds are needed for planting one hectare area (i.e. approx.150 - 200 g seeds/acre).

(Note: 1 hectare/ha = 10, 000 square meter)

Some growers use hybrid seeds for growing tomatoes; however, cost of growing is higher in such cases.

Nursery-raised seedlings may be transplanted when they reach 10-15 cm in height. Spacing for the winter crop is 75cm x 60 cm and for the spring-summer crop, recommended spacing is 75 cm x 45 cm.

Manures and Fertilizers: Tomato plants prefer fertile soils. Therefore, 20 to 25 tons FYM (farmyard manure) or compost or any other organic manure per hectare should be incorporated in the soil at the time of land preparation.

(Note: 1 ton = 1000 kg)

Standard NPK fertilizers in multiple doses may be applied for the healthy growth of the pants. 275 - 300 kg of ammonium sulphate/ha may be applied as top dressing one month after planting. Foliar application of 35 - 40 kg of nitrogen and 45 - 50 kg phosphate per hectare in four or five sprays may be beneficial for the crop. Concentration of foliar spray should be less than 1%. Concentrations higher than 1% may scorch the leaves.

Irrigation: Need-based watering is done for tomato plants; however, both overwatering and insufficient irrigation should be avoided. Young plants need frequent watering until they get established in the fields. During summers, irrigation is needed at weekly intervals. During winters, irrigation may be done at fortnightly intervals. Mulching is a good practice to conserve soil moisture. Mulching also suppresses weed growth.

Companion Cropping: Tomato plants can successfully be grown as companion plants along with carrots, parsley, dill, mints, dandelions, asparagus and marigolds.

Training and Pruning: Tomato plants may need training and pruning in some cases. In case of training, single stem training is practiced where all young shoots are removed from the main stem. The main stem grows vigorously utilizing all resources available.

Weed Management: Weed control is an important practice; generally manual weed control is practiced. In case of severe weed infestations, chemical herbicides may be used.

Disease Management of Tomato Plants: Tomato plants are affected by a number of diseases. A list of these diseases is given below:

Damping off: Damping-off is a fungal disease caused by a group of fungi including *Phytophthora spp.*, *Pythium spp., Rhizoctonia spp.* and *Fusarium spp.* and it mainly affects seedlings and young plants. Using poor quality infested seeds, improper planting depth, high salt concentrations in the soil, poor soil drainage resulting in damp, wet seed beds and fields, and nutrient deficiencies in the soil are some of the reasons for the prevalence of damping off disease. Major symptoms include collapse of the seedlings and root rot of the seedlings and young plants. Best control measures include using high quality, pathogen-free seeds and planting materials, providing good soil drainage, and maintaining proper soil pH levels. Spraying of the soil with a copper based fungicide such as Bordeaux mixture may reduce the incidences of damping off disease up to a great extent.

Fusarium Wilt: This is a fungal disease caused by *Fusarium oxysporum*. Fusarium wilt is a soil-borne fungal disease and is characterised by wilting of the plant. The disease first affects the leaves and the affected leaves curl upward and inward, and become yellow and finally fall off from the plant. Since it is a soil-borne disease, the best control measures include soil treatment with a recommended fungicide or soil sterilization before planting the crop. Planting tomato varieties that are resistant to Fusarium wilt disease may also be tried. Spraying the soil with copper fungicides like Bordeaux mixture may prevent the spread of the disease up to a great extent.

Late Blight or Phytophthora Blight: This is a fungal disease caused by *Phytophthora spp*. This disease attacks the leaves and stems of the plants as well as the fruits. The disease is prevalent in warm, humid climate and in wet waterlogged areas. Major symptoms include leaf blight, fruit rot, and root rot and the affected plants wilt and die in due course of time. Control measures include avoiding overwatering the plants and providing proper soil drainage. Spraying of the soil with a copper based fungicide such as Bordeaux mixture may reduce the incidences of this fungal disease up to a great extent.

Leaf Mold: It is caused by the fungus *Cladosporium fulvum*. There will be yellowish green blotches on the upper surface of the leaves accompanied by grey or greenish brown moldy spots. The disease is prevalent in warm, humid climate. Proper cultural practices that ensure proper aeration of the plants may control the disease up to some extent. In case of severe infestations, a standard chemical fungicide may be used; however it is not advised due to the health risks associated with chemical-residues.

Bacterial Wilt: Bacterial Wilt is a seed-borne, bacterial disease caused by *Pseudomonas solanacearum*. Leaves of the affected plant wilt first and eventually the entire plant wilt and die. The best control measures include the use of disease-free seeds and planting materials and adopting the cultural practices such as crop rotation and removal of the diseased plants from the field as soon as possible.

Leaf Curl Virus: This is a serious viral disease of tomato plants and is spread by white flies. The leaves of affected plants start curling and malformed. Best control measures include controlling white flies, observing good field sanitation practises, planting virus-resistant varieties and removal of the infected plants as soon as possible. However, complete control of viral diseases is often not possible.

Tobacco Mosaic Virus (TMV): Tobacco Mosaic Virus (TMV) is a highly infectious viral disease of the tomato plants. There will be chlorotic areas on the leaves of the affected plants. The disease is spread by vectors such as aphids and other insects that infest the plants. Major symptoms include curling leaves, mottled fruits, stunted plant growth and excessive leaf drop. Best control measures include controlling vectors such as aphids and other insects, observing good field sanitation practises, planting virus-resistant varieties and removal of the infected plants as soon as possible. However, complete control of viral diseases is often not possible.

Early Blight: This is a fungal disease caused by *Alternaria solani*. It affects the foliage and causes brown spots on immature fruits. Control measures are similar to those of late blight disease.

Bacterial Canker: This is a bacterial disease caused by *Corynebacterium spp*. There will be cankers and light brown or dark streaks inside stems. Control measures are similar to those of bacterial wilt disease.

Anthracnose: Anthracnose is a fungal disease caused by the fungi *Colletotrichum piperatum* and *C. capsici*. The disease is prevalent in warm and humid climate and mainly affects the fruits. Sunken, circular spots of 2-3 cm diameter appear on affected fruits and in severe cases, a mass of pink or yellow fungal spores appear on fruits. Best control measures include use of disease-free seeds and seed treatment with a recommended fungicide before sowing. Crop rotation and field sanitation may help prevent the spread of the disease up to a great extent.

Cercospora Leaf Spot: Cercospora Leaf Spot (Frog Eye) is a seed-borne fungal disease caused by *Cercospora capsici*. Major symptom is small brown circular, watery leaf lesions followed by leaf drop. Control measures include use of disease-free seeds and planting materials, and adoption of best cultural practices such as crop rotation and regular field sanitation. In case of severe infestations, a standard chemical fungicide may be used; however it is not advised due to chemical-residue risks.

Grey Mold: Grey Mold is a fungal disease caused by the fungus *Botrytis cinerea*. It affects young shoots and flowers and affected plant parts collapse completely and a mass of grey-coloured fungal spores appear on the surface of the dead plant parts. The disease is prevalent in warm, humid climate. Proper cultural practices that ensure proper aeration of the plants may control the disease up to some extent. In case of severe infestations, a standard chemical fungicide may be used; however it is not advised due to chemical-residue risks.

Insect-Pest Management of Tomatoes: Major insect-pests of the tomato plants include tomato caterpillars, Epilachna beetles, aphids, white flies, spiders, thrips, jassids, and red mites. Root-knot nematodes may also pose a serious problem in certain soils. A detailed account of various insect-pest management strategies for tomato plants is given below:

Tomato Fruit Worms or Fruit Caterpillars: These are small caterpillars that eat the leaves and feeds on the vegetative parts of the plants. They can be controlled effectively by handpicking or by spraying organic pesticides based on pyrethrums.

Epilachna Beetles: These beetles are capable causing serious losses in a tomato crop. Both larvae and adults feed on the young leaves and tender shoots of the plants. In case of milder infestations, hand picking of the beetles and destroying them may be advised. Spraying the plants with organic pesticides such as diluted soap solution or neem emulsion or organic pyrethrum-based pesticides may control these beetles up to a great extent.

Root Knot Nematodes: Root Knot Nematodes are small microscopic worms that penetrate the roots of the plants and cause swollen root nodes or gall formation in the roots. The affected plants are unable to absorb water and vital plant nutrients from the soil and eventually, they wilt and die. Root knot nematodes are a common problem in sandy soils and therefore adding organic matter to sandy soils to increase its fertility and water-retention ability may reduce the problem up to some extent. Crop rotation may also be tried to reduce root knot nematode problem. So far the best control measure for root knot nematodes is to plant nematode-resistant varieties of tomato plants.

Thrips and Aphids: Thrips are small, slender-looking, sucking insects that may be white, yellow, brown or black in colour. They attack tender leaves and shoots and the affected leaves become stunted and distorted in appearance and curl upwards. Aphids are also small sucking insects that suck cell sap from tender foliage and plants. Both thrips and aphids may be controlled by using natural predators like ladybugs or by planting trap crops like marigolds. Spraying the plants with organic pesticides such as diluted soap solution or neem emulsion or organic pyrethrum-based pesticides may control these insects up to a great extent. In case of severe infestations, a standard chemical insecticide may be used; however it is not advised due to chemical-residue risks.

Flea Beetles, Pepper Weevils, and Pepper Maggots: Flea beetles are about 2mm long, shiny black insects and feed on the seedlings and young plants leaving holes in them. Pepper weevils are about 3mm long black insects and feed on the young and mature plants. Pepper maggots are whitish yellow 1cm long maggots that feed on the core of the fruits. In case of milder infestations, hand picking of the insects and destroying them may be advised. Spraying the plants with organic pesticides such as diluted soap solution or neem emulsion or organic pyrethrum-based pesticides may control these insects up to a great extent. In case of severe infestations, a standard chemical insecticide may be used; however it is not advised due to chemical-residue risks.

Spider Mites: Spider mites are very small, pale yellow insects that feed on the underside of leaves and the infected leaves gradually curl downwards. In severe infestations, leaves become speckled with a webby appearance. Spraying plants with a strong flow of water may get rid of these insects up to some extent. In case of severe infestations, a standard chemical insecticide may be used; however it is not advised due to chemical-residue risks.

White Flies: Whiteflies are tiny insects of 1.5mm long that suck cell sap from the leaves. Eventually the affected leaves turn yellow and fall down. White flies may be controlled by following good cultural practices, such as removing and destroying infected plants, and field sanitation and by using color traps or by using organic pesticidal solutions such as spraying soap solution, neem emulsions, or pyrethrum-based insecticides.

Integrated Pest and Disease Management for Tomato Plants: IPM (integrated pest management) may be practiced for pest management. IPM comprises of cultural control such as crop rotation, hand picking of caterpillars and beetles, and using trap crops; mechanical control such as using yellow colour cards and pheromone traps as insect baits; biological control such as using ladybugs for controlling aphids; and if necessary, chemical control by using recommended chemicals.

IDM (integrated disease management) may be used for controlling diseases; this is similar to IPM. IDM uses a combination of cultural, mechanical, biological, and chemical control measures for managing crop diseases.

Physiological Disorders of Tomatoes: Tomato plants and fruits are susceptible to different types of physiological disorders. A detailed account of most popular disorders of tomato plants is given below:

Sunscald or Sunburn of Fruits: Tomato fruits that are exposed to sunlight develop whitish or yellowish patches on fruit skins. This disorder is most prevalent in hot and summer days. Fruits may be protected from sunscald by covering them up with grass or similar materials.

Puffiness or Puffy Fruits: Puffed weightless fruits of the tomatoes are also a common disorder found in tomatoes. This is mainly caused by the poor pollination of the flowers.

Blossom-End Rot (BER): This disorder is characterised by the rotting of the blossom end or base of the fruit. It is caused by calcium deficiency in the fruits. Water stress, high temperature, high humidity, root damage or any stress condition that interferes with Calcium-uptake may cause BER. Regular foliar sprays of calcium chloride (36% calcium) or calcium nitrate (20% calcium) may prevent the incidence of this disorder.

Green Shoulder: This disorder is characterised by the presence of a permanent dark green area at the top end of the ripening fruit. It is a genetic disorder but its occurrence can be minimized by providing proper aeration to the plants during warm periods.

Blotchy Ripening or Gray Wall: This disorder is characterised by the presence of blotchy, gray areas on green fruit. Even if the rest of the fruit starts ripening and turns red, these blotchy areas remain gray. Most likely, this disorder is caused by wide fluctuations in the environmental conditions such as high moisture and humidity, and low light and temperature. Using varieties that are resistant to blotchy ripening disorder may be the best way to tackle this problem.

Catface: This disorder is characterised by the cracking of the fruit at the blossom end. This is caused by the temperature fluctuations during fruit set. Using varieties that are resistant to Catface disorder may be the best way to tackle this problem.

Zippering: This disorder is characterised by the presence vertical cracks in the form of a zip. Major reason for zippering is wide variations in the growing environment of the plants. Regulating growing environmental conditions to the optimum level may solve this problem effectively.

Minute Cracking: This disorder is characterised by the presence of minute, often inconspicuous cracks along the shoulder of the fruits. Major reason for minute cracking is wide variations in the growing environment of the plants. Regulating growing environmental conditions to the optimum level may solve this problem effectively.

Radial Cracking: This disorder is characterised by the presence of numerous visible cracks that originate from the top end of the fruit and proceeds downwards in a radial fashion. Major reason for radial cracking is wide variations in the growing environment of the plants. Regulating growing environmental conditions to the optimum level may solve this problem effectively.

Concentric Cracking: This disorder is characterised by the presence of concentric cracks around the calyx or top end of the fruit. Major reason for concentric cracking is wide variations in the growing environment of the plants. Regulating growing environmental conditions to the optimum level may solve this problem effectively.

Fruit Splitting: This disorder is characterised by the splitting of the fruit. It happens when fruits are exposed to high temperatures or due to the wide variations in day and night temperatures. Growing plants in optimum temperature conditions will solve this problem effectively.

Harvesting: Tomatoes can be harvested at different maturity stages depending upon the purpose of uses. Immature green and mature green tomatoes are harvested for raw uses. Half-ripe and red ripe tomatoes are harvested for culinary and processing purposes. For shipping and export purposes, firm mature green fruits are harvested. Tomatoes for canning are harvested when they are fully ripe.

Packing: Freshly harvested tomatoes may be packed in wooden boxes, bamboo baskets, gunny bags or similar materials.

Figure 13: Bamboo Baskets for Packing Tomatoes

Yield: Yield varies from 20 to 25 tons per hectare. Yield of hybrid varieties varies from 50 to 60 tons per hectare.

Figure 14: Green, Red, Yellow and Orange Tomatoes

Storage: Optimum storage temperature is from 12°C to 15°C. When stored at freezing-point, the fruits show chilling injury. Mature green tomatoes can be kept for as long as 14-15 days at 10°C to 15°C. Ripe tomatoes can be kept for 10 days at 5°C. Optimum relative humidity for storage is 90-95%.

Nutrient Deficiency Symptoms: Tomato plants are highly sensitive to the nutrient deficiencies in the growing medium. Therefore it is very essential that the plants should be fed with all essential nutrients required for its healthy growth. Major nutrients required for the plant growth are Nitrogen (N), Sulfur (S), Phosphorous (P), Potassium (K), Calcium (Ca), Magnesium (Mg), Iron (Fe), Boron (Bo), Zinc (Zn), Manganese (Mn), Copper (Cu), and Molybdenum (Mb). A detailed account of various functions of these nutrients is given in the table below:

Nutrient	Function
Nitrogen (N)	Protein synthesis and chlorophyll synthesis
Sulfur (S)	Protein synthesis
Phosphorous (P)	Synthesis of proteins, phospholipids, sugar phosphates, and nucleic acids
Potassium (K)	K helps formation of carbohydrates and proteins; and helps in transpiration regulation and photosynthesis
Calcium (Ca)	An important constituent for the formation of cell wall
Magnesium (Mg)	Chlorophyll synthesis
Iron (Fe)	Fe plays an active role in respiratory process, and chlorophyll synthesis
Boron (Bo)	Bo role plays an active in carbohydrate breakdown
Zinc (Zn)	Zn plays an active role in carbohydrate metabolism, CO_2 utilization, and phosphorus metabolism
Manganese (Mn)	Mn plays an active role as an activator of several enzymes of aerobic respiration
Copper (Cu)	Cu is a chief constituent of ascorbic acid oxidase system and helps in achieving Carbon/Nitrogen balance in plants
Molybdenum (Mb)	Mb plays an active role in nitrogen fixation and nitrate reduction

If these nutrients are not present in the growing medium, plants may suffer from various nutrient deficiencies. A detailed account of various nutrient deficiency symptoms in tomato plants is given in the table below:

Nutrient	Deficiency Symptoms
Nitrogen (N)	Chlorosis (chlorophyll deficiency in plants) and stunted plants. Excess nitrogen leads to excessive vegetative growth and therefore delayed fruiting and ripening processes
Sulfur (S)	Small leaves, pale green leaves, and suppressed fruit formation
Phosphorous (P)	Premature leaf fall, poor fruiting, reduced copper and zinc availability
Potassium (K)	Chlorosis, stunted plants with numerous tillers, and little or no flowering
Calcium (Ca)	Stunted plant growth and brown spots along the leaf margins
Magnesium (Mg)	Chlorosis
Iron (Fe)	Chlorosis in a mottled pattern
Boron (Bo)	Stunted root growth; stem elongation, sterile flowers or lack of flowering, deformed fruits, and die-back of stems
Zinc (Zn)	Intervenal Chlorosis and reduced availability of Fe
Manganese (Mn)	Chlorosis and reduced availability of Fe
Copper (Cu)	Reduced vegetative and reproductive growth
Molybdenum (Mb)	Necrosis of leaf tissues

Post Harvest Practices for Tomatoes

In order to increase the shelf life of tomatoes, the fruits should be harvested at correct harvest maturity, i.e. when fruits are well ripe and are just turning colour with a pinkish tinge. Seeds should be well-developed and fruit should be well-filled with juice. Good quality fruits are characterised by uniform shape, uniform colour and smooth skin and are free from pest-disease infestations, and all defects including cracks, bruises and mechanical injuries.

Precooling: Precooling of freshly harvested produce is essential to remove field heat from the produce and also for maximising postharvest keeping quality. Recommended temperature for precooling is 12.5°C. Both room cooling and forced air cooling may be practiced.

Optimum Storage Conditions for Tomatoes: Depending upon the harvest stage, storage temperature requirements vary. For 'Mature Green' tomatoes optimum storage temperature is 12.5-15°C. They can be stored up to 14-15 days at this temperature without any quality loss. Beyond 14 days storage, tomatoes may likely to start decaying. For 'Light Red' tomatoes storage temperature should be 10-12.5°C. They can be stored for 8-10 days at this temperature without any decay. 'Firm-Ripe' tomatoes can be stored at 5-10°C for 10 days without any quality loss.

Tomatoes are sensitive to low temperatures below 10°C and exposing the produce below 10°C may result in chilling injury.

Ripening: Ripening process in tomatoes is increased at high temperature and high humidity environment (i.e. 12.5-25°C and 90-95% R.H.) For ripening process, tomatoes can be kept in a well-ventilated ripening room at 18-21°C with 90-95% R.H. (relative humidity). Ripening room should have good aeration or ventilation to prevent carbon di oxide accumulation. It is found that tomatoes stored at 20°C (68°F) at 90-95% relative humidity have maximum retention of vitamin C and optimum coloration with their nutritional quality intact.

For products in transit, tomatoes may be kept at 14-16°C for slowing down the ripening process. For uniform coloring, tomatoes may be treated with 100 ppm ethylene for 2-3 days.

90-95% relative humidity is essential to maximize produce quality and prevent water loss from fruits. Long periods of exposure of fruits to high humid environment may result in rotting and spoilage of the produce.

Organic Growing Practices for Tomatoes

Growing plants organically is becoming a healthy trend in today's world. Because of the eco-friendly growing practices adopted in organic agriculture, the products of organic agriculture are extremely safe to consume without fearing any dangers and health risks associated with pesticidal residues in foods.

In organic farming practices, growers use planting materials of organic origin only. They adopt IPM (integrated pest management) strategies for controlling diseases and pests of the crop plants. No synthetic chemicals are used to control pests and diseases of the plants. Growers use only organic manures and organic fertilizers for plant nutrition. IPM as such heavily relies on various cultural, mechanical and biological practices for disease-pest management. IPM emphasises on using a lot of beneficial cultural practices such as crop rotation, companion planting, using trap crops etc for maintaining a healthy ecosystem in the crop-growing fields.

So in a nutshell, we can say that THREE important points to be considered while growing an organic crop are:

1. Choosing organic planting materials
2. Preparing soil and enhancing soil fertility by using organic methods
3. Integrated Pest-Disease Management Practices

Choosing Plant Materials: In organic production, we need to use planting materials of organic origin. Organic planting materials may be raised in grower's own nurseries or may be purchased from a certified organic plant nursery.

Preparing Soil: For growing any plants, as we all know, soil is the major growing medium. So soil should be fertile with all essential plant nutrients to support a healthy plant growth. A soil analysis may be done to determine the fertility status of the soil. According to the soil test results, various measures may be taken to enhance soil fertility. In organic farming, soil fertility is enhanced by the addition of organic manures and organic fertilizers to the top soil before planting the crops.

Soil Sterilization: In organic farming, seedlings are planted in a sterilized soil medium so that soil-borne plant pathogens can be prevented. Soil sterilization can be done by soil solarization method. Soil solarization is a method of trapping solar energy within the soil by covering the soil with a transparent polyethylene cover for a certain period of time. This practice kills all soil-borne plant pathogens. After soil sterilization, methods need to be adopted to enhance soil fertility. This is accomplished through the addition of organic manures and organic fertilizers, incorporating green manuring crops to the top soil, and by using various soil conditioners and amendments.

Types of Organic Manures and Organic Fertilizers: While growing plants organically, we generally use organic manures and fertilizers as well as biofertilizers. Any organic matter that may be used as an organic fertilizer is known as organic manure. Manures contribute to the fertility of the soil by adding organic matter and nutrients. There are three main classes of organic manures used in soil management. These are, *animal manures; compost* and *plant manures or green manuring crops.*

How to Apply Fertilizers? We all know that under-nourishment of plants may lead to poor growth and development. But overnourishment is also not good for plants. Over-fertilization or adding extra fertilizer doses to plants may lead to nutrient toxicity. There are five types of nutrient toxicity. These are:

1. Chlorosis (yellowing of plant tissue caused by a shortage of chlorophyll synthesis)
2. Necrosis (death of plant tissues)
3. Accumulation of anthocyanins (production of a purple or reddish colorization of foliage and/or stems)
4. Lack of new growth (stunting or reduced growth)
5. Stunted new growth

So while applying fertilizers, certain rules are to be observed. These rules are:

1. Apply right quantities of fertilizers
2. Apply fertilizer when the plants can best use the nutrients
3. Apply small amounts of fertilizer frequently
4. Be careful not to over fertilize the plants

Pest and Disease Management in Organic Farming: IPM or Integrated Pest Management practices may be adopted for insect-pest management in organic gardens and organic farms. Similarly IDM or Integrated Disease Management practices may be used to control plant diseases. IWM or Integrated Weed Management practices may be adopted for controlling crop weeds. Integrated pest and disease management practices make use of all available eco-friendly practices to control pest and disease infestations in the fields.

Major IPM Components: There are FOUR components of IPM practices. These are:

1. Cultural Control : This method makes use of all available cultural practices such as trap crop technology, crop

rotation, mulching, use of cover crops, companion planting, intercropping etc
2. Biological Control: This method makes use of natural enemies or predators of insect-pests and plant pathogens. In other words, biological control practices make use of all available biopesticides and biocontrol agents for ensuring plant protection
3. Mechanical Control: This method makes use of all mechanical devices that help in controlling crop pests and diseases; for example, light traps, poison baits, colour traps etc
4. Chemical Control: In severe cases of pest and disease infestations, growers may use those chemicals which are permitted in organic growing practices as per the country's organic production standards

Growing Practices for Organic Tomato: Tomato crop require low-medium rainfall conditions for growth. Winter growing has been found ideal for organic tomato cultivation. Well drained sandy loam soils (which are naturally fertile) may be chosen for growing organic tomatoes. Acidity of the soil has to be checked regularly at the start of each season. Extremely acidic soils are to be avoided. If it is not possible, correct the pH level of the soils before growing tomatoes. Tomato plants prefer slightly acidic soils to neutral soils. Optimum pH for tomatoes is 6.0 to 7.0.

Crop Rotation: Crop rotation is an important cultural practice in growing organic tomatoes. Crop rotation should be done with non-solanaceous crops like legumes, and pulses. Cropping systems like okra-tomato and tomato-onion are quite popular in tropical regions like India. Crops like rice, cauliflower, watermelon, garlic, cotton, sunflower etc can also be grown after tomato. At least a gap of one year should be maintained between two plantings of Solanaceous crops (brinjal, capsicum, chili, potato, tomato etc).

Buffer Zone or Isolation Distance: It is very important to create a buffer zone around an organic farm in order to avoid any contamination from nearby conventional farms. For growing organic tomatoes, a buffer zone of 7.5-15 meters is recommended.

Field Preparation: The field should be ploughed to reach a fine texture, after which FYM (farm yard manure)@10-15 tons/acre or vermicompost@1-1.5 tons/acre should be incorporated into the top soil. Beds are raised after preparing the field. Drainage channels are made at 50 cm breadth for every 1 meter of the bed.

Propagation: Tomato plants are generally propagated through seeds. Organic cultivation requires the use of certified organic seeds. The selected seeds should also be of disease resistant, high yielding and pest resistant. The seeds can be purchased from certified organic plant nurseries.

Seed Treatment: Pre-sowing treatment of seeds with Trichoderma culture@1gm per 150 gm of seeds builds up its resistance to pests and diseases.

Tomato Varieties Suitable for Organic Farming: Not all tomato varieties are suitable for organic farming. For organic farming of tomato, the open pollinated varieties are preferred. Indian varieties like Swarna Lalima and Swarna Naveen are suitable for organic cultivation in the tropics. Lakshmi NP 5005 is another popular variety with resistance to bacterial wilt and leaf mosaic virus. A list of tomato varieties that are resistant to various diseases is given below:

1. Varieties resistance to Bacterial Wilt: Arka Abha, Swarna Lalima, Arka Alok, Arka Abhijit (hybrid)
2. Varieties resistance to Leaf curl: Anmol, Aditi (hybrid), Kashi Amrut, Avinash-2 (hybrid)
3. Varieties resistance to Blight: Akash, Vajra, Meghana

4. Varieties resistance to Verticillium and Fusarium wilt: Empire, Rupali (hybrid), Roma
5. Varieties resistance to Nematodes: SL-120, Ronita

Raising a Nursery: Nursery beds of 1m x 3m specification at 20 cm height are to be made. It has been calculated that 12-15 such beds would suffice to raise seedlings (or for sowing 150 g seeds) for planting in 1 acre area. Standard application of organic manures to the soil, followed by sterilization by solarization reduces pest incidence during early stages. The seeds need to be planted at intervals of 5 cm with 2 cm spacing between successive seeds planted. The depth of planting should not exceed 1 cm. The covering of beds after sowing, with nylon sheets prevent pest infestation in seedlings. Seeds start germinating within 7-14 days. Germination is best in warm temperatures of 20-24° C.

Planting: Seedlings of 20-25 days can be used for transplanting. The spacing followed for transplanting in the main field is 60 cm between rows and 50 cm between plants. The planting should not be too dense to prevent further growth of plants, nor too light for weeds to grow. The irrigation and drainage channels must be distanced from plants by 30 cm.

Weed Management: Weed control is an important cultural activity to follow particularly 4-5 weeks after transplanting. Use of mulches, crop rotation, sanitation, shallow tilling etc need to be followed to prevent the growth and spread of the weeds. Regular weeding is necessary particularly when the seedlings are small.

Figure 15: Mulching of Tomato Plants with Dried Leaves and Straw

Staking: The tomato plants grow well with staking in the form of small bamboo sticks, branches or wires for support. This should be done 15-20 days after the transplantation or when the seedlings attain 20 cm height. This will help with their branching and will also increase fruit production.

Figure 16: Staking of Tomato Plants

Manuring and Watering: 10 – 15 tons of FYM or 1-1.5 tons of vermicompost per acre is the standard application rate at the time of field preparation. Treatment of FYM with Trichoderma@500 gm/load of FYM manure is highly recommended.

Watering or irrigation depends on the soil type, and climatic conditions. Generally, the field need to be irrigated once in 7 days.

Crop Rotation for Pest Management: Crop rotation is one of the traditional methods used to break the life cycle of pests and insects. Careful monitoring of the field and crop is essential to identify the presence of pests. It has been observed that crop rotation is fairly successful in pest control for organic tomatoes.

Trap Crops for Pest Management: Trap crops are also used for preventive pest control. Sweet corn is a trap crop used to attract the tomato fruit worms. Marigold is used to attract the tomato fruit borers that will otherwise cause serious crop loss in tomato fields. For every 16 rows of tomato, one row of marigold should be sown as a trap crop. The marigold seedlings must be 15 days older to the tomato seedlings. This is to facilitate synchronized flowering of the crops.

Biological Control for Pest Management: Biological control of pests has been found to be the most effective pest management strategies in case of organic tomato farming. The use of Trichogramma is effective against Lepidopteran pests of tomatoes. Neem-based pesticides are effective against the fruit borers of tomatoes. Trichoderma is suitable for seed treatment and effective against root fungi. Neem seed extract and cow dung mixture is also used to prevent excessive flower drop in the tomato plants.

How to Control Fruit Borers? Control measures for Fruit borer (*Helicoverpa armigera*) include the following measures:

- Look out for Helicoverpa eggs on the top leaves and hand pick and destroy the larvae
- Use of resistant varieties like Rupali, Roma etc.
- Usage of American marigold as a trap crop
- Border cropping using sorghum (8 rows) at 30cm x 10 cm spacing support natural predators like Coccinellids and Chrysoperla
- Attracting predatory birds by using bird perches or any other luring techniques
- Neem seed kernel 5% spray to stop the pests at its early stages
- Severe borer attack can be controlled by application of seed extracts of Strychnosnux-vomica to the soil, at the rate of 1.5 g/plant at every 20 days (twice)
- Use of biocontrol agents like *Bacillus thuringiensis*@1g/liter of water) or *Trichogramma chilonis*@50,000 eggs, six times at weekly intervals
- Setting up of pheromone traps (15 traps/ha) with Helilure, changing every fortnight

How to Control Serpentine Leaf Miners? Control measures for Serpentine leaf miner (*Liriomyzatrifolii*) include the following measures:

- Intercropping with field beans (1 row for every 8 rows of tomato), with fields beans sown 12 days before tomato transplanting
- Phyto-sanitation
- Neem seed kernel extract 5% spray or ginger, garlic, chili extract (1litre/tank)

How to Control Tobacco Caterpillars? Control measures for Tobacco caterpillar (*Spodoptera litura*) include the following measures:

- Exposure of soil to extreme heat through ploughing (before transplanting)
- Flooding of field while field preparation, in order to kill hibernating larvae
- Use of castor plant as a trap crop (125 plants/ha) will attract egg laying moths; Eggs of the pest can be collected from the castor plant and destroyed
- Laying pheromone traps (15 traps/ha) with pheroclin SL lure for monitoring pest population
- Neem seed kernel 5% spray for protection against larvae

How to Control White Flies? Control measures for White fly (*Bemisia tabaci*) include the following measures:

- Use of nylon net in nursery to protect the seedlings
- Phyto-sanitation
- Irrigation control
- Use of natural predators, *Brumus* and *Chrysoperla*
- Use of pearl millet as a barrier crop around the main field, it should be sown 15 days prior to the transplanting of the main crop

- Removal of weed hosts
- Use of yellow sticky traps (50 traps/ha)
- Neem seed kernel extract 5% spray

General Pest Management Practices for Organic Tomato: A general remedy for pest management is the spraying of decoction made of leaves from aloe vera, neem, Ocimum tenuiflorum (basil), Achyranthes aspera and Aristolochia bracteata. The boiled decoction is mixed with water (100 ml/litre of water) and sprayed on the plants.

Disease Management in Organic Tomatoes: Tomato plant is attacked by a number of diseases caused by various pathogens such as fungi, bacteria, and viruses. There are some physiological disorders also found in the tomato plants, most of which are caused by Abiotic factors and environmental stresses. Two major physiological disorders of tomato are: *cat face* and *blossom end rot*.

The major diseases that attack the root system of the crop include fusarium wilt, verticillium wilt, bacterial wilt, and rhizoctonia wilt. Diseases like early blight, leaf spot, bacterial canker and late blight attack above-ground stems and foliage. Diseases like bacterial spot, bacterial speck and anthracnose affect the tomato fruits.

How to Control Damping Off? Control measures for Damping off (*Pythium aphanidermatum*) include the following measures:

- Use of certified seeds
- Sterilization of soil by solarization
- Better drainage facilities
- Neem cake (400 g/sq. m) application to nursery bed 15 days prior to sowing and irrigation at 4 days interval
- Use of light, draining soil for nursery beds; light and frequent irrigation

- Use of well decomposed organic manures
- Seed treatment with leaf extract of *Bougainvillea glabra* (20ml/liter of water) for 6 hours

How to Control Early Blight? Control measures for Early blight (*Alternaria solani*) include the following measures:

- Selective crop rotation
- Distancing tomato and potato cultivation areas
- Sanitation of crop by removing and burning of infested branches, leaves
- Spray of 5% eucalyptus or lantana leaf extract during evening hours
- Treatment with *Trichoderma viride* or *Pseudomonas fluorescens* (5g/100g of seeds)

How to Control Fusarium Wilt? Control measures for Fusarium wilt (*Fusarium oxysporum* f.*lycopersici*) include the following measures:

- Selective crop rotation
- Seedling root tip treatment in solution of turmeric and asafetida (10 g each/liter of water) prior to transplanting
- Sanitation of field
- Use of resistant varieties
- Spraying of 15 days old diluted *Panchakavya* biodynamic solution (1:10)

How to Control Powdery Mildew? Control measures for Powdery mildew (*Leveillula taurica* and *Erysiphe polygoni*) include the following measures:

- Spraying of solution of milk and water (1:1) once in 3 days after disease occurrence

How to Control Bacterial Canker? Control measures for Bacterial canker (*Clavibacter michiganense*) include the following measures:

- Crop rotation by avoiding subsequent rotation of solanaceous crops
- Spraying of cow dung extract
- Hot water seed treatment

How to Control Bacterial Wilt? Control measures for Bacterial wilt (*Pseudomonas solanacearum*) include the following measures:

- Crop rotation with field beans, maize, soybean or cruciferous vegetables
- Seedling treatment (root dip) in asafetida solution (10 gm/liter of water)

How to Control Leaf Curl Virus? Control measures for Leaf curl (Gemini virus) include the following measures:

- Crop rotation
- Use of healthy seedlings
- Soil sterilization
- Crop sanitation
- Spray of 5% Neem seed kernel extract to control white flies (vectors of Gemini virus)

General Disease Management Practices for Organic Tomato: One general disease preventive measure to be taken up is crop rotation with corn, cereals and sorghum. Field sanitation and proper disposal of affected plants is also essential. Covering of exposed fruits with straw will help prevent sun scald (blotches and dry skin in fruits).

Harvesting: The crop takes 2-3 months for reaching maturity. Harvesting stage is determined by market requirements. The stages of fruit harvest are immature green, mature green, turning pink, half ripe, red ripe and over ripe. Turning pink fruits just show color at the blossom end while mature green are not pink. Half ripe fruits are majorly covered in pink color. The mature green fruits are best suited for export purposes. For domestic fresh consumption, the turning pink or half ripe stage is the best. For the purpose of seed production, the red ripe tomatoes are best. Canning and processing stages could use the red and over ripe stage fruits as per requirements.

Yield: It is estimated that yield of an organic tomato crop varies from 10-15 tons/acre, under irrigated conditions. Yield may be slightly lesser than that of a conventional crop.

Hydroponic Growing Practices for Tomatoes

Tomatoes are suitable for growing in hydroponics growing systems. Hydroponics is a practice of growing plants in a nutrient-rich solution, without soil.

Today the term 'hydroponics' has become synonymous with 'soil-less production of crop plants' though the term itself means that it is water ('hydro' means 'water') that is at work ('ponics' means 'labor' or 'work'). The term 'hydroponics' may seem to represent a sophisticated process but in reality, hydroponics is a simple process of crop production. The only difference of a hydroponics crop production from that of a traditional method is that hydroponics makes use of a nutrient solution as plant growing medium instead of soil.

Hydroponics under controlled environmental conditions (i.e. greenhouse hydroponics) is more successful than that of outdoor hydroponics. When higher yields per unit area and higher productivity per plant are desired, greenhouse hydroponics is preferred than the outdoor hydroponics.

Hydroponics is based on the principle that plant growth in a traditional soil-based production system is not dependent on the soil rather it is dependent on the nutrients and moisture present in the soil. So, if the plant nutrients and moisture required for the plant growth are provided through any other medium other than the soil, plants can still have a natural growth. Therefore in a hydroponics system, ideal nutrient and moisture requirements of the plants are fulfilled through a water culture or solution culture under ideal environmental conditions. In other words, hydroponics is *soil-less crop production system under controlled environmental conditions*.

There are mainly two types of hydroponic grow systems: Aggregate systems and Deep Water Culture (DWC) systems.

Figure 17: Hydroponic Tomatoes

Aggregate or Substrate System of Hydroponics: These systems use aggregate growing media such as rock wool, perlite, coir, peat moss, etc as plant root support system. For growing tomatoes, rock wool or peat moss may be used as substrates. However, DWC hydroponics, particularly NFT systems are preferred for commercial hydroponics of tomatoes.

DWC Hydroponics: These systems use only water culture or nutrient solution to grow the plants. These systems do not use any aggregate growing media as plant root support system. The plant roots are totally suspended in the nutrient solution. A tray made of plastic or Styrofoam boards or similar materials that float on the surface of the solution is used to support the plant above the solution. Holes are provided on the tray so that the roots are inserted into the solution while the shoots stand on the tray growing upwards. In DWC systems, the nutrient solution needs to be aerated or bubbled continuously by using an air pump and air stones. Nutrient solution should be changed regularly and kept at constant level in the reservoir tank. Best examples of popular DWC systems are Nutrient Film Technology (NFT) and BubblePonics. NFT is one of the most popular water culture hydroponics systems used for growing vegetables. Bubble hydroponics is water culture hydroponics where constant oxygenation or aeration of the plant root zones is required for the healthy growth of the plants.

NFT Hydroponics of Tomatoes: A standard NFT hydroponics grow system is divided into TWO areas: a propagation area (propagation room) and a grow area (grow room). Propagation area is where propagation takes place while grow area is exclusively dedicated for vegetative growing purposes.

A standard NFT hydroponics system includes grow trays, reservoir tanks, plant pots, plant nutrient kits , pH test kit, digital TDS meter, water pumps, air pumps and air stones, air diffusers, growing media kits, starter materials (seeds, clones), and a plant starter tray. The system should have air filters, duct mufflers or silencers, and a light proof system .Safety of operations is ensured in a hi-tech hydroponics system by providing provisions for fire protection, and insulation.

Propagation: It is advised to start planting materials in the hydroponics system itself. This helps avoid transplant stress for the plants. Propagation within the hydroponics system will also help the growers obtain disease-free and pest-free starting materials. A separate propagation room may be used for propagation purposes.

Figure 18: Tomato Seedlings

Propagation room should be specifically designed for facilitating propagation process by providing optimum temperature, light and humidity conditions. A standard propagation system available with the hi-tech hydroponics system includes a starter tray, a humidity dome, a seedling heat mat, a heat mat thermostat and a growing media kit. A starter tray is used for seed germination. Seedlings are grown here until they attain transplanting age. Starter trays are made up of durable plastic and come with a reservoir tank in which nutrient solution is prepared to be supplied to the seedlings. A humidity dome is used for providing optimum humidity conditions for accelerating germination process. Heating the root zone of germinating seeds is important for achieving their optimum growth. Seedling heat mat is used for this purpose in a hydroponics system. A heat mat thermostat may be used to regulate the temperature of the heat mats.

In tomatoes, propagation is by seeds. Seeds are sown directly on the starter/seed tray that is filled with a growing medium. After sowing, tray is covered with a lid. The growing medium is constantly kept moist by pumping the nutrient solution kept in a reservoir tank. Rockwool is the most popular growing media used in the hydroponics. Rockwool cubes with a hole in the center are popularly used for seed propagation. These cubes are soaked in water or nutrient solution before placing seeds into the hole. Cubes are kept moist until the germination process is completed and seedlings are ready for transplanting. ***Seeds germinate within 5-10 days. 4-6 weeks old seedlings (15-20 cm tall seedlings) may be transplanted in the grow areas.***

Seedlings, when they are ready for transplanting, are taken to the grow room. A grow room has a perfectly balanced environment with proper ventilation, humidity, temperature and light management system.

A controlled environment such as a greenhouse is recommended for growing tomatoes because such a system provides shelter, and stress-free environment for the plants. Temperature of the growing environment may be monitored regularly by using a thermometer. ***Temperature within the growing environment should be kept at 60- 80° F.***

A hygrometer may be used to measure the humidity inside the growing environment. A well-designed hydroponics system has cross air flow system to ensure adequate aeration around the plants and their root zones.

Preparation of Nutrient Solution: For healthy plant growth, a plant needs both macronutrients and micronutrients (trace elements). Major macronutrients include Nitrogen (N), Phosphorous (P), Potassium (K), Calcium (Ca), Magnesium (Mg), and Sulphur(S). Trace elements are Iron (Fe), Manganese (Mn), Boron (Bo), Zinc (Zn), Copper (Cu), and Molybdenum (Mb). Nutrient formulas containing all these nutrients in correct proportions are available in the market as hydroponic nutrient mixes. A grower may purchase them to prepare the nutrient solution for the hydroponics.

Regarding, preparation of the nutrient solution, care should be taken that only good quality water is used. Hard water should be avoided by all means. Periodical flushing of the nutrient solution is necessary to prevent salt build up in the solution.

While preparing an ideal nutrient solution, pH, electrical conductivity (EC), temperature and total dissolved solids (TDS) of the solution should be measured as each of these parameters has an impact on the degree of nutrient absorption by the plant roots.

Optimum pH range for the nutrient solution should be 5.5-6.5. Large variations in the nutrient pH may lead to poor absorption of nutrients by the plants.

pH of the nutrient solution can have a great impact on the plant growth. Since every plant has a preferred pH range at which plant nutrients become available to its growth, solutions having too low or too high pH should be avoided in a hydroponics system. pH of nutrient solution should be checked regularly by using any of the pH devices available in the market. Less expensive pH devices such as a pH control kit and pH pen may serve this purpose for those who are looking for cost effectiveness. A pH meter may be a costly device as compared to a pH control kit but provides instant reading.

Electrical Conductivity (EC) of nutrient solution should be between 2.0 and 5.0. EC refers to 'electrical conductivity' or flow of electric current through the nutrient solution. EC and concentration of the nutrient solution is proportionately correlated. i.e. when the concentration of nutrients is higher in the solution, EC will be higher and vice versa. EC meter is used to measure the electrical conductivity of the nutrient solution. EC meter records the reading in either micromhs per centimeter (uMho/cm) or microsiemens per centimeter (uS/cm).

The temperature of the nutrient solution affects the reading of the EC meter. Hence it is recommended that EC should be measured at 25^0 C always. If the temperature of the nutrient solution is above 25^0 C, the EC reading will be higher, even though concentration of the solution remains same. If the temperature of the nutrient solution is below 25^0 C, EC reading will be on the lower side.

Total Dissolved Solids (TDS) of the nutrient solution should be between 1400 and 3500 ppm. TDS refers to the total dissolved solids present in the nutrient solution. A TDS meter is used to measure TDS level of the nutrient solution. The meter reading is shown in parts per million (ppm).

In hydroponics, the roots of the growing plants should be aerated at regular intervals because plant roots need oxygen in order to survive. This oxygenation of roots may be carried out by using highly efficient air pumps and air stones.

Remember, good aeration of the hydroponic solution is essential to obtain the best results.

If nutrient solution is poorly aerated, it adversely affects the root development. A healthy root system is white coloured and highly branched. In poorly aerated solution, roots develop browning and turn to dark colour. Regular monitoring of the nutrient solution provides the grower an idea about the status of the nutrient solution.

Regular testing of the nutrient solution for pH, EC, and TDS helps growers ensure that plants are being fed with right nutrients at right concentration. It also helps monitor the salt levels of the nutrient solution at every phase of crop production. Thus growers can take appropriate corrective measures if the salt levels rise unexpectedly resulting in a 'salt build-up' in the growing system. Two important corrective measures recommended for eliminating problems associated with salt build-up is regular flushing of the growing medium with a fresh nutrient solution or replacing the nutrient solution with a fresh one.

Remember, temperature of the nutrient solution should always be lower than the air temperature.

Tips for Successful Nutrient Application

1. Whenever a nutrient solution is prepared, use a measuring cup to take correct quantities of nutrients
2. Make sure that nutrients are taken in correct proportions
3. If nutrients are in powder form, use warm water to dissolve them and mix well by stirring vigorously to get a homogenous solution
4. Use a PPM (parts per million) measuring device (e.g.: nutrient monitor) to measure the concentration level of the nutrient solution

There should be an air filtration system in the growing system. Carbon filters effectively eliminates undesirable odours from the grow system. Tomatoes need plenty of light. ***Adequate lighting should also be provided for the growing plants@8-10 hours of light/day***. Fluorescent grow lights or LED lights may be used for providing artificial light.

LED lights are highly energy efficient and economical. LEDs are low temperature way to increase the amount of light that plants receive. Light distribution and coverage within the system can be adjusted by installing panels and reflectors.

Placement of the lights should be directly related to the intensity of the light required by the plant. If more light intensity is required, place the light close to the plants but not too close to burn the leaves. Adjustable lighting system may be used to adjust the light according to the plant requirement.

In hydroponics, care should be taken not to expose the roots of the growing plants to the light. Root exposure to light may induce growth of algae and thus contaminate the growing medium.

CO_2 forms an integral part of a plant growth system and therefore it is important that CO_2 should be applied in a hydroponics system for healthy plant growth. Generally CO_2 is administered to the plants through a tank application process.

Significance of Light Energy and CO_2 in Hydroponics: Plants need light energy for various purposes, major being photosynthesis and transpiration. During photosynthesis, plants produce carbohydrates (foods) using light energy, carbon dioxide and water. In an enclosed hydroponics system, artificial lighting system and CO_2 application system may be used to provide the light and CO_2 needed by the plants.

Pest and Disease Management: In indoor hydroponics in greenhouses, pests and disease can also be effectively controlled by protecting the greenhouses from the entry of insects and pests. Wire meshing may be used at all entry points to prevent pest infestation. If proper hygiene is practiced within the greenhouses, disease incidences can be minimised to nil.

Planting and Care: 4 seedlings may be planted per square meter area. Staking may be provided for growing plants. Vigorously growing plants may be pruned and trimmed to keep them in shape and encourage flower production.

Fruit Production: 45-60 days after transplanting seedlings, fruit set begins. Fruit production may be encouraged by exposing the plants to longer durations of light per day.

Yield: In well-managed hydroponic grow systems, a grower can produce a good quality crop within 60-75days. Average yield of fruits is estimated to be 4-5 kg/m^2. Both quality and quantity of the crop can be optimised in a well-managed hydroponic grow system.

Pollination in Tomato Flowers: When tomatoes are grown under enclosed and controlled environments, hand pollination of flowers may be necessary to induce fruit set.

Figure 19: Tomato Flowers

Greenhouse Growing Practices for Tomatoes

  Tomato plants are suitable for greenhouse gardening also. The most advanced greenhouse technology for growing tomatoes involves the use of high-tech greenhouses with substrate hydroponic grow systems with built in irrigation systems. These greenhouses are equipped with sacks of substrate (rockwool, peat moss or coconut fiber or perlite) for growing hydroponic tomatoes within greenhouses.

A greenhouse is a specially designed structure covered with a transparent material (either polyethylene sheets or glass) and is provided with environmental control systems.

There are two types of greenhouses: polyhouse that use polyethylene as covering material and glasshouse that use glass as covering materials. In recent years, Polytunnel greenhouse growing of tomatoes has been becoming very common for commercial-scale production.

Figure 20: Greenhouse Tomatoes

Indeterminate types of tomatoes are mostly suitable for greenhouse production. One common method of greenhouse growing is to plant the seedlings directly on the substrates or directly on the ground. Growing seedlings require staking and an individual plant may reach above 2.5 meter in height under good management conditions.

A greenhouse is meant for growing plants under protected environment that is established through a series of environmental control systems. Greenhouse cover acts as its protection boundaries which allow only sunlight to enter and no other external environmental factors are allowed inside while maintaining a fully healthy ecosystem within the greenhouse that is suitable for plant growth. Thus, a greenhouse provides right environment for the plant growth during all seasons. Major objective of a green house is to provide optimum environmental conditions for optimum crop production at maximum yield and productivity throughout the year.

Significance of Regulating Light in Greenhouses: Plant height and flowering are controlled by light intensity and light duration. Plant growth in response to light duration is termed as photoperiodism. Plants are categorized into three groups based on their light requirement: short day plants, long day plants and day neutral plants. Short day plants require short duration of light exposure while long day plants require long duration of light exposure. In day neutral plants, light intensity and light duration has no effect on the growth and flower production of the plants. In greenhouses, light can be controlled as per the plant requirements.

Light Management in Greenhouses: Greenhouse cover filters harmful ultra violet radiation. Sufficient light can be provided by using artificial lights also. One of the advantages of using artificial lighting system in greenhouses is that, light intensity and light duration can be adjusted according to the seasonal changes and plant requirements. Light can be extended in case of long day plants and reduced in case of short day plants. In case of shade-loving crops, additional light proof material can be used to insulate greenhouse cover.

<u>Note: Tomatoes are day-neutral plants.</u>

Significance of Regulating Air in Greenhouses: Oxygen and Carbon dioxide are two important gases that play an important role in plant growth. Plants manufacture their foods in the presence of sunlight during day time through a process called photosynthesis. During photosynthesis, plants use carbon di oxide and release oxygen. However, during night, plants release carbon di oxide. So it is important that oxygen-carbon di oxide ratio inside the greenhouses should be suitable for plant food production.

Controlling Air Movement in Greenhouses: Air Mixers are used for the air movements within the greenhouses. They help to create a uniform environment congenial for plant growth within the greenhouses. This will also helps in preventing the incidences of plant diseases.

Significance of Regulating Humidity in Greenhouses: Plant growth is affected by the presence of water vapour in the air. So it is essential that relative humidity within the greenhouses is kept at optimum level always.

Relative Humidity for Tomato Plants: Relative humidity of 60-70% is ideal for greenhouse tomatoes.

Significance of Regulating Temperature in Greenhouses: Temperature also affects photosynthesis and respiration. Thermotropism is nothing but the changes in plant response according to the variations in day and night temperatures.

Temperature Management in Greenhouses: Evaporative cooling pads, electric fans, and sensors may be used within the greenhouses for temperature management. Evaporative cooling pads reduce temperatures up to 15^0 C during the dry period. Electric Fans with air replacement capacity and cooling operation are used for climate control. Indoor sensors are used for the temperature and humidity control within the greenhouses. Outdoor sensors are used for detecting external temperatures, humidity, wind speed and wind directions.

Heating systems may also be used to regulate temperatures within the greenhouses. Major heating systems used for the greenhouses are steam, hot water, forced air heaters and infrared radiant heaters.

Greenhouse Temperature for Tomato Plants: A day temperature of 70-80^{0}F (20-25^{0}C) and night temperature of 60-64^{0}F(15-20^{0}C) is ideal for greenhouse tomatoes.

Suitable Tomato Varieties for Greenhouse Production: There are hundreds of tomato varieties available for greenhouse cultivation. The most popular among them are Vendor, Caruso, Tropic, Medallion, Electra, Perfecto, Capello, Gabriela, Laura and Trust.

Containers for Growing Greenhouse Tomatoes: Seed trays having 36 or 48 or 72 cells may be used for sowing seeds. Seeds germinate within 7-10 days. 4-6 week old seedlings may be transplanted. Earthen pots, rock wool slabs, polybags or plant beds may be used for growing plants. Peat moss and sand-based growing medium may be prepared to fill the containers and plant beds. Such a growing medium will ensure proper aeration and drainage.

Individual plants should be planted in individual containers. If plants are grown in plant beds, proper spacing should be followed. A spacing of 1m between two rows and 50 cm between two plants may be ideal in most of the cases. Recommended planting density for greenhouse tomatoes is approx. 10,000 plants/acre.

Figure 21: Greenhouse Tomato Plants on Plant Beds

Irrigation Systems: Spaghetti tube irrigation system that carries irrigation water by a ¾ inch polyethylene tube to each pot may be used for potted plants. Another method is drip irrigation where water is carried through drip lines. It is very expensive but has highly efficient use of water.

Figure 22: Irrigation of Greenhouse Tomatoes

Sprinkler irrigation or misting may also be tried. In this irrigation method, water is supplied overhead by spray nozzles. It is effective if plants are raised in plant beds prepared within the greenhouses.

Fertilizer Application: Fertigation may be tried for greenhouse tomatoes. Application of fertilizers through irrigational water is called fertigation. Fertigation system allows adjustments of fertilizer application program according to both growth stage and growth rate of the plants and changing climate conditions.

All types of fertilizers such as organic, inorganic and bio-fertilizers may be used for greenhouse production. However, using organic and biofertilizers are considered as more safe and ecofriendly way of growing plants. Examples of organic fertilizers are compost, vermicompost, bone meal and farm yard manure.

Biofertilizers are cultures of naturally occurring microorganisms that are capable of fixing nutrients in the growing media. For example, N-fixing rhizobial culture and azolla fungal culture.

Note: It is better to avoid the application of inorganic fertilizers or chemical fertilizers within the greenhouses.

Pruning: Single stem pruning is recommended for greenhouse tomatoes. In single stem pruning, only the main stem is allowed to grow while removing all the lateral shoots or 'suckers'. Single stem pruning is essential for producing large-sized good quality fruits. Suckers or lateral shoots should be removed at regular intervals either at weekly or fortnightly intervals.

Training: Training of the tomato plants is an important cultural operation in greenhouse cultivation. Vines of the growing plants may be trained on trellises or wire meshes. Some growers use wooden stakes also for supporting the plant growth.

Pollination: Proper pollination of female flowers is necessary for fruit production. Tomato flowers are 'perfect' flowers containing both male and female parts. In field tomatoes, flowers are often pollinated by wind. However, in greenhouses and such enclosed spaces, flowers should be pollinated either by hands (hand pollination) or by a machine (electric pollinator). During pollination, optimum environment with 70-80^{0}F (20-25^{0}C) and 70% relative humidity should be kept within the greenhouses. Regular pollination of flowers at least twice a week is important for maximum fruit production. Fertilization normally happens after two days of pollination. Fertilized ovary gradually develops into a fruit.

Harvesting and Yield: In greenhouses, tomato fruits may be harvested 3-4 months after sowing seeds. Yield depends upon the growing practices and varietal selection. On average, the yield of a greenhouse crop of beefsteak or the cluster tomatoes is approx. 50 kg/m^2.

Advantages of Greenhouse Production: Green House technology provides a controlled and favorable environment for crop growth. Environments can be suitably modified as per the requirements of the crop. Greenhouse crops yield high irrespective of seasonal changes. Greenhouse helps to grow plants throughout the year. Greenhouse allows clean cultivation and incidences of pests and diseases are very less. If insect proof nets are installed at the entry points of the greenhouses, it prevents insect-pest infestations from outside. In greenhouses, Carbon dioxide released by the plants during the night is consumed by the same plants themselves during the day time. Thus the plants get about 8-10% times more food than the open field conditions. Off-season production and round-the-year crop production is possible by the manipulation of growing seasons.

Productivity and quality of fruits are also good in case of greenhouse crops. Higher production per unit area, i.e. high crop yield and higher quality crops may be achieved through greenhouse production.

Greenhouse allows optimum and efficient use of fertilizers and pesticides through fertigation. It is easy to train vining plants in greenhouses by providing staking or training the trailing stems along the trellises.

IPM or Integrated Pest Management in Greenhouses: Use insect proof nets at all entry points and doors for preventing entry of insect-pests from outside. If insects or pests are present within the greenhouses, use manual or biological control measures for insect-pest management. For example, hand-picking of insects and destroying them, and spraying the plants with neem-based emulsions or biopesticides. Use insecticides or pesticides only if it is necessary. Use manual, cultural or biological methods of control for weed management such as hand-hoeing or hand-pulling of weeds. Use sterilized growing medium or soil for plant growth. Inspect or regularly monitor greenhouse environment for possible infestation of insects and pests. Inspect plants regularly for any disease outbreak. Adjust environmental control systems to prevent pest and disease infestations.

Nutrient Solutions for Greenhouse Tomato Production: For maximizing the productivity of the greenhouse crop, properly mixed nutrient solutions should be used. Ready-made nutrient mixes are available in the market. Growers may use these fertilizer mixes or they can prepare their own nutrient solutions. However while feeding the plants, one should remember that the plants have different nutritional requirements at different levels of growth stages. So a fertilizer formula used for young plants may not be useful for mature plants.

In other words, any crop nutrient management program requires a thorough understanding of the nutrient solution concentrations in parts per million (ppm) for the various nutrients required by the plants. A grower should know how to manage the concentrations of individual nutrients required by the plant so that he or she can maximize the growth and productivity of the crop. As mentioned before, plants may be fed either by using premixed products available in the market or by the grower- formulated nutrient solutions.

Note: Elaborating on various commercially available hydroponic or greenhouse nutrient mixes is beyond the scope of this book and hence readers are suggested to consult their local plant nutrient suppliers for further guidance on this topic.

Figure 23: Greenhouse Tomatoes

Container Gardening Practices For Tomatoes

The practice of growing plants in pots and containers to raise a garden is known as container gardening. Container gardening is most suitable for urban homes, and multistorey buildings of towns and cities. Tomatoes, being a productive crop, are suitable for growing in containers.

Figure 24: Container Gardening of Tomatoes

Before raising a container garden of tomatoes, the following points need to be addressed.

1. *Select a Tomato Variety that is Suitable for Container Gardening:* Tomato plants with a bushy growth are the most suitable varieties for growing in containers

2. *Select Suitable Containers/Pots:* Containers of any shapes and sizes may be used as per the requirement of the grower
3. *Select Suitable Growing Media:* Potting mixture should be fertile and ideal for growing tomatoes

Suitable Containers for Container Gardening of Tomatoes: There are lots of choices available in the market. A detailed account of various containers available for gardening purposes is given below:

- Earthen pots and pans: Earthenware pots are most commonly used containers for growing vegetables like tomatoes
- Wooden barrels and planters: These containers should be painted from inside as well as outside with waterproof oils-paints
- Plastic jars, pots, dishes and bowls: They are best suited for growing small plants
- Glazed clay and china (porcelain) pots, shallow bowls and troughs and pottery containers in contemporary designs : These are best suited for indoor gardening
- Other Containers: Boxes and crates, polybags, cement pots (suitable for large plants), cans and buckets, tin boxes, drums, and brass and copper containers (these are not suitable for warm climate)

Containers may be of any shape such as circular, round, oval, elliptical, cone-shaped, pyramid-shaped, rectangular or square shaped and heart-shaped containers.

Figure 25: Tomatoes Grown in Polybags

However there are certain rules to be followed while choosing an appropriate container for growing tomatoes. These are:
1. Adequate drainage: Containers should have at least one hole of an adequate size at the bottom as in earthen pots, to drain out excess water
2. Containers can easily be portable
3. Containers can hold sufficient volume of growing medium
4. Containers should be lightweight and easy to handle
5. Containers should be durable and free of toxic substances
6. Containers should prevent root circling

Growing Medium or Potting Mixture for Container Gardens:
Growing media may be prepared from a mixture of good soil; river sand; well-decomposed organic manure (compost or farmyard manure) ; nitrogenous fertilizers (urea or ammonium sulphate) and small amounts of recommended insecticides and fungicides. An ideal growing medium or potting mixture will have the following characteristics:
1. Growing medium should be able to hold the seedlings firmly
2. It should be free of eggs and larvae of insect-pests and disease-causing pathogens
3. It should have good water-holding capacity
4. It should have excellent aeration and drainage

How to Prepare an Ideal Growing Medium? Mix good soil, river-sand and well-rotten organic manure (cow dung, compost, vermicompost etc) in equal quantities with the help of a *khurpi* or shovel. Make sure that the mixture is free from various soil-borne insects, termites, red ants and cut worms. Add a small quantity of recommended fungicide to the mixture before filling it in the containers; this helps to prevent seedling rotting caused due to fungal infections. After raising a crop for one season, the container mixture should be removed and cleaned of roots and exposed to the sun for a few days. This growing medium could then be reused after mixing one-third the quantity of organic manure and a small quantity of recommended fungicide.

Importance of a Garden Calendar: Preparing a garden calendar is a useful gardening practice. Garden calendar allows a gardener to record important dates and events such sowing time, fertilizer schedule, watering schedule, weeding and aftercare of the plants.

Garden Tools: A container gardener needs to have all necessary garden tools such as spade or shovel, watering can, small hand-sprayer, bamboo stakes and strings, measuring tapes, duster, garden knives and scissors, garden baskets etc

Tomato varieties suitable for container gardening in India: Tomato varieties such as Pusa Early Dwarf, Pusa Gaurav, Pusa Hybrid No. 1 and Pusa Hybrid No. 2 are suitable for container gardening in India and other tropical regions.

Sowing Seeds: Seeds may be sown directly in the containers. Alternatively, seedlings may be raised in a nursery to be transplanted in the containers later. Seeds may be sown either during Jan.-Feb or during Sept.-Oct. seeds germinate with one or two weeks. 60-65 days after seed sowing, first crop of tomato fruits may be harvested.

Propagation and Care of the Tomato Plants: General Rules
- Seedlings may be raised either by sowing the seeds directly in containers or in well-prepared nursery beds
- When seedlings are raised in nursery beds, transplanting of seedlings in the suitable containers should be done at the right time; normally one or two week old seedlings are transplanted
- While transplanting, a single healthy seedling may be transplanted in each container. Sometimes two or more seedlings may be transplanted in each container
- In direct seeding, 4–5 seeds may be sown in each container; later these seedlings may be thinned out by keeping only one or two healthy plants per container
- Plants in pots and containers need a lot of care and attention
- It is essential to water frequently depending on the season, size of the plant and container
- Plants need extra water in dry summer season, so watering should be done twice a day (morning and evening)
- Too much watering can be as harmful in winter as too little in summer
- In the rainy season, proper water drainage is essential
- If there is heavy rain, containers should be tilted slightly to drain out the excess water from the top

Fertilizer Application and Plant Nutrition: Topdressing with nitrogenous fertilizers improves plant growth and yield. Urea or ammonium sulphate may be applied frequently in small quantities during the active vegetative growth of the phase. In general, 5–10 gm of urea may be applied in moist soil once a week or 10 days, starting from 3 weeks after sowing or 2 weeks after transplanting. High dose of fertilizer is very harmful since it can kill the plants. If urea or ammonium sulphate is applied in dry soil, the plants must be watered immediately.

Staking: Growing tomato plants need staking. Bamboo stakes or wooden stakes may be used to provide support for the growing plants.

Weed Management: Hand-hoeing and weeding with the help of a small khurpi or spade should be done periodically to remove weeds. Weeds should be uprooted gently by hand without disturbing the roots of the growing plants.

Insect Pest Management: Major insects found in container-grown tomato plants are aphids and jassids, white flies, fruit flies and fruit borers. Aphids and jassids are small-sucking insects, injuring the plants especially in early stage of their growth. Fruit flies and fruit borers damage young fruits and make them unfit for consumption. Use of organic insecticides such pyrethrum-based sprays or tobacco emulsions, or neem-oil based solutions will effectively control these insects. Use of mechanical traps (colour traps, light traps etc) and manual picking of insects may also be tried for insect control. After spraying with insecticides, vegetables should not be harvested for 7 days for consumption in order to avoid any possible health-risks associated with insecticidal residues.

Disease Management: Fungal diseases (damping off and wilt) and viral diseases affect the tomato plants particularly during the rainy season. Fungal diseases can be controlled by drenching the soil with an appropriate fungicide. Virus affected plants should be removed and destroyed.

Harvesting: Tomatoes are harvested at various ripening stages as per consumer requirements. Generally, tomatoes are allowed to ripen on plants before harvesting.

Watering: Determine water needs of the plants by weighing pots and feeling the soil or growing medium. Using some indicator plants that readily show water stress alongside tomato plants is also a good idea. As a general rule, watering should be done in early morning or late evening to minimize evaporation loss. Applying water in two or more applications conserves water.

Nutrition and Health Benefits of Tomatoes

Tomatoes are one of the most widely consumed vegetables across the entire world. Tomatoes are consumed raw as well as cooked. Both ripe and unripe fruits are consumed. Tomatoes are considered as one of the most nutrient-dense plant-based foods. A detailed account of nutrient composition of raw and cooked tomatoes is given in the following tables.

Nutrition in Raw Fresh Tomatoes: A detailed account of nutrients present in 100g of edible portion of **raw tomatoes** is given below:

	Nutrient	Unit	Green Tomato	Orange Tomato	Yellow Tomato
			Value/100g		
1	Protein	g	1.2	1.16	0.98
2	Fiber	g	1.1	0.9	0.7
3	Calcium	mg	13	5	11
4	Iron	mg	0.51	0.47	0.49
5	Potassium	mg	204	212	258
6	Zinc	mg	0.07	0.14	0.28
7	Vitamin C	mg	23.4	16	9
8	Thiamin	mg	0.06	0.046	0.041
9	Riboflavin	mg	0.04	0.034	0.047
10	Niacin	mg	0.5	0.593	1.179
11	Vitamin B-6	mg	0.081	0.06	0.056
12	Vit. B-12	mg	0	0	0
13	Folate	µg	9	29	30
14	Vitamin A	IU	642	1496	n/a
15	Vitamin E	mg	0.38	n/a	n/a
16	Vitamin K	µg	10.1	n/a	n/a
17	Vitamin D	µg	0	0	0

Nutrition in Cooked Tomatoes: Cooked tomatoes are a low-calorie food with good amounts of protein and dietary fiber. A detailed account of the nutrient composition of **cooked red tomatoes** (as per USDA nutrient database) is given in the table below:

Nutrient	Unit	Value per 100 g
Water	g	94.34
Energy	kcal	18
Protein	g	0.95
Total lipid (fat)	g	0.11
Carbohydrate	g	4.01
Fiber, total dietary	g	0.7
Sugars, total	g	2.49
Calcium, Ca	mg	11
Iron, Fe	mg	0.68
Magnesium, Mg	mg	9
Phosphorus, P	mg	28
Potassium, K	mg	218
Sodium, Na	mg	11
Zinc, Zn	mg	0.14
Vitamin C	mg	22.8
Thiamin	mg	0.036
Riboflavin	mg	0.022
Niacin	mg	0.532
Vitamin B-6	mg	0.079
Folate, DFE	µg	13
Vitamin B-12	µg	0
Vitamin A, IU	IU	489
Vitamin E (alpha-tocopherol)	mg	0.56
Vitamin K (phylloquinone)	µg	2.8

Health Benefits of Tomatoes: Tomatoes are an excellent source of dietary fiber, antioxidants, vitamins and minerals. A detailed account of various health benefits that tomatoes offer is given below:

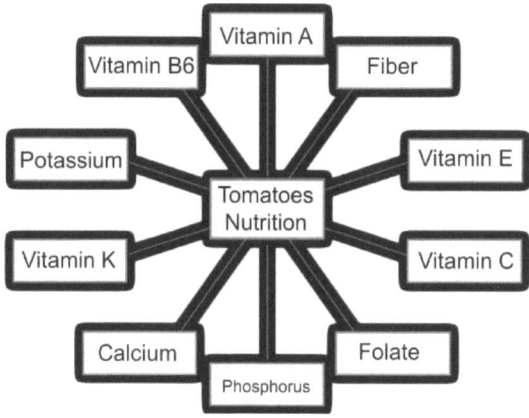

Figure 26: Nutrients in Tomato

Tomatoes are Rich in Antioxidants: As we know, tomatoes are an excellent source of lycopene and other antioxidants. Antioxidants are capable of reducing the free-radical damage of the cells of our body and thus help us prevent many lifestyle diseases such as cancer, heart diseases, and obesity. Researches reveal that cooking tomatoes will actually increase the antioxidant activity of this food, though some amount of water-soluble vitamins such as vitamin C may be lost during cooking process.

Tomatoes are Rich in Dietary Fiber: Dietary fiber is good for human body. Tomatoes are rich source of dietary fiber. High fiber foods are good for weight loss as they take long time to get digested and also make you feel full for a long period of time. High fiber food aids in digestion, cures constipation, lowers blood cholesterol, cleanses the gastrointestinal tract and may reduce the risk of developing diabetes and colorectal cancer.

Tomatoes are Rich in Calcium (Ca): Calcium is the most abundant mineral found in the human body. Major functions of Ca are blood clotting, nerve impulse and muscle contraction, and protection of bones from thinning. Calcium promotes bone health and teeth health. Calcium deficiency results in weakening of bones, rickets in children, tooth decay and pains in legs and back of the body.

Tomatoes are Rich in Phosphorus (P): Phosphorus is an essential component for the formation of high energy compounds and various nucleic acids. Phosphorous is a major component of bone and teeth and it also increases body's immunity. It also plays an important role in the formation of nerve cells. Phosphorous deficiency may result in anaemic appearances, weaknesses in muscles, poor immune symptoms etc. Phosphorus is second most abundant mineral, after calcium, in the human body.

Tomatoes are Rich in Potassium (K): Potassium is an essential mineral that plays an important role in the growth of human body. It regulates body's fluid balance and also helps in lowering blood pressure. Potassium is an essential mineral required by the body for the proper functioning of heart. Potassium deficiency may lead to irregular heartbeat, kidney failure, insomnia, lung failure, and general weakness of the human body.

Tomatoes are Rich in Vitamin A: Vitamin A is also known as Retinol. It is essential for eye health. It also strengthens body's natural immune system. Vitamin A is also essential for tissue building, and skin health. Vitamin A deficiency results in night blindness, and drying of skin and eyes.

Tomatoes are Rich in Vitamin B6: Vitamin B6 is also known as Pyridoxine. It is essential for fat metabolism and protein metabolism. It also helps in the production of RBCs and neurotransmitters. Vitamin B6 facilitates proper functioning of estrogen and testosterone hormones in the body. Deficiency symptoms include depression, improper functioning of immune system and sores in mouth.

Tomatoes are Rich in Vitamin B9 or Folate: It is also called Folic acid or Folate. It is essential for energy production from food. It helps in synthesis of nucleic acids and proper functioning of immune system and blood production by facilitating functioning of iron and increasing production of RBCs. It also helps in controlling amino acid metabolism.

Major deficiency symptoms include birth defects in new born babies, diarrhoea, hearing loss due to ageing, improper functioning of immune system, weakness, fatigue and headaches. Regular consumption of folic acid helps in slowing down progression of hearing loss with ageing; to prevent birth related defects in new born babies; for protection from cancer, heart diseases, depression and degeneration of body due to ageing; and to prevent memory loss and osteoporosis.

Tomatoes are Rich in Vitamin C: It is also known as ascorbic acid. It is a powerful antioxidant vitamin. Vitamin C helps in absorption of iron and calcium. It increases body's natural immunity. Vitamin C deficiency results in a disease called scurvy. Major symptoms of scurvy are bleeding gum, joint pain, and hair loss.

Tomatoes are Rich in Vitamin E: Vitamin E is essential for strengthening body's natural immune system and cardiovascular system. It is a powerful antioxidant vitamin and hence protects the body from heart diseases and cancer. Vitamin E deficiency results in weakening of muscular system and nervous system. Other deficiency symptoms include lack of coordination and balance.

Tomatoes are Rich in Vitamin K: Vitamin K is essential for blood clotting, and for preventing heart diseases, cancer, and osteoporosis. Vitamin K deficiency results in bleeding gums and bleeding nose.

Figure 27: Food Uses of Tomatoes

Food Uses of Tomatoes

There is no dearth of the variety of foods that can be prepared from tomatoes. Raw ripe tomatoes are used for making a wide variety of fresh salads and fresh juices. Tomatoes are cooked in several ways. Cooked tomatoes are used for preparing many kinds of vegetable dishes and processed products. Tomatoes may be boiled, baked, roasted, stuffed, and fried for cooking purposes. Tomatoes may be preserved for long term usage by drying or dehydrating or by salting. Tomatoes are cooked and processed in different ways to make numerous types of delicious food preparations. A detailed account of some popular tomato food dishes is given below:

Tomato Salads: Fresh ripe red tomatoes are an essential ingredient in many types of vegetable salads. Tomatoes may alone be used for making salads or used as an ingredient along with other salad vegetables such as lettuce, cucumbers and onions for making fresh salads.

Figure 28: Tomato Salad

Boiled Tomatoes: Tomatoes may be blanched and boiled for easy peeling and pulping process. Tomato pulp is used for making soups, sauces, puree etc.

Figure 29: Boiled and Peeled Tomatoes

Stuffed Tomatoes: Stuffed tomatoes are delicious food delicacies that are served with main meals. Large size firm ripe tomatoes are used for making this food preparation. Different types of fillings both vegetarian and non-vegetarian may be used for stuffing.

Figure 30: Stuffed Tomatoes

Grilled Tomatoes: Grilled and barbecued tomatoes are often served with other barbequed foods.

Figure 31: Grilled Tomatoes

Tomatoes in Fast Foods: Tomatoes are an essential ingredient in many fast food items such as burgers, noodles, pizza, sandwiches etc.

Figure 32: Burger with Tomato Slices

Tomato Pizza: Tomato slices, tomato sauce, and tomato rings are used as major ingredient in pizza preparations. Can you imagine a pizza without tomato toppings?

Figure 33: Tomato Pizza

Tomato Sandwiches: Similarly, we cannot imagine sandwiches without tomatoes. Today tomatoes have become an essential ingredient in sandwich preparations.

Figure 34: Tomato Sandwich

Fried Green Tomatoes: Sliced green tomatoes coated with corn flour and then fried are used as snacks in many countries.

Tomato and Egg Preparations: There are several ways of cooking tomatoes with eggs. One of the easiest breakfast preparations is '*Stir-Fried Tomato and Scrambled Eggs*.' Sliccd tomatoes are stir-fried in a few drops of oil and then eggs are added to make this simple breakfast recipe. Finely chopped tomatoes are added to eggs to make *'Tomato Omelette,'* another breakfast preparation.

Stewed Tomatoes: Sliced tomatoes with fresh herbs are used for making tomato stews. Stewed tomatoes are used like any other vegetable stews.

Figure 35: Tomato Stew

Tomatoes in Mixed Vegetable Preparations: Tomatoes are used for making different types of vegetable and pulse preparations. In Southern India, *tomato sambar* is a popular vegetable dish made with tomatoes and pulses or *dal*. Similarly, *tomato rasam* is another popular tomato dish.

Figure 36: A Tomato Vegetable Dish

Tomatoes in Rice, Pasta and Noodle Preparations: Use of tomatoes is not limited as fast food ingredients or vegetable ingredients. Tomatoes are extensively used in various types of spicy rice preparations such as tomato rice, and *biriyani,* and noodle preparations. Tomatoes are added in various types of pasta and pastry preparations also.

Figure 37: Tomato Rice

Processed Tomato Foods

A large number of processed foods are prepared from tomatoes. Tomatoes are used for making various processed foods such as purees, paste, sauces, jams, cocktails, marmalade, pickles and chutneys. Tomatoes are canned for using as an ingredient in various processed products. Major among the tomato-based processed foods are tomato juice, tomato puree and paste, tomato sauce or ketchup, tomato chutney, tomato cocktail, tomato soup, canned tomatoes, tomato powder, and tomato pickles. A detailed account of various preserved and processed products of tomato is given below:

Sun-Dried Tomatoes: Solar drying or oven drying is the best way to preserve excess tomatoes for long term usage. Dried tomatoes are used just like fresh tomatoes for making tomato pastes and purees. Before using them, just soak them in water.

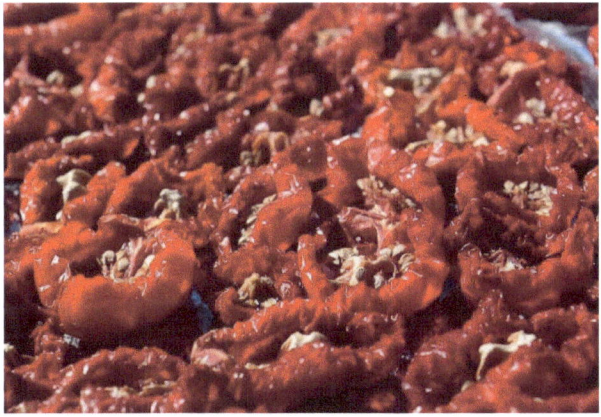

Figure 38: Dried Tomatoes

Tomato Jam: Good quality ripe red tomatoes are cleaned and chopped finely before mixing with sugar and lime juice. This mixture is cooked along with red chilli flakes, and assorted spices and herbs on a low flame until jam like consistency is obtained. Tomato jam is prepared and used just like any other fruit jam.

Figure 39: Tomato Jam

Tomato Juice: Plant-ripened or fully-ripe red, fleshy fruits are selected for juice preparation. Clean tomatoes are then sorted based on their quality. Only good quality juicy tomatoes are selected and trimmed before slicing them into 5-6 pieces. Sliced tomatoes are taken in a big jar and heated at 70-90 degree Celsius for 3-5 minutes to facilitate pulping process.

Pulping Process: Pulping or extraction of juice may be done either manually by using a stainless steel sieve (recommended for small quantities) or by using a pulping machine (pulper or spiral press) for commercial-scale tomato juice production. Juice is sieved by using a strainer to remove seeds and other particles. A good quality juice should be of deep red colour and must contain about 0.4

percent acid; it should be uniform in appearance and should have high nutritive value.

Preservation of Tomato Juice: For every one liter of tomato juice, add 10 grams of sugar, 5 grams of salt, 1 gram of citric acid and 1 gram of sodium benzoate. Heat this juice mixture at 82-88 degree Celsius for a minute for homogenization, i.e. to make the consistency of juice uniform. Hot tomato juice is filled into cans and sealed. Sealed cans are then sterilized by immersing them in boiling water for 30 minutes. Sterilized cans are then cooled and stored at ambient temperature in dry place.

Figure 40: Tomato Juice

Tomato Soup: A simple tomato soup may be prepared in two steps by using the following recipe. First step is spicing the tomato juice and the second step is cooking process.

Spicing the tomato juice: Ingredients Required
- 1 Kg of fresh tomato juice
- Spice bag: Make a small cloth bundle of 20 grams of chopped onion, 5 grams of chopped garlic, 1 gram of large cardamom powder, 1 gram of cinnamon powder, 1 gram

of black pepper powder, 1 gram of cumin powder and Clove 5 numbers. (Note: Use clean ,dry white muslin cloth)

Process: Strained tomato juice is taken in a vessel and is brought to boil. The boiling juice is stirred continuously to make it a thick concentrate. All the while, spice bag is dipped into the mixture. For the next step, take 350 milliliters of water, 20 grams of salt, 20 grams of sugar, 20 grams of butter or milk cream, and 10 grams of corn flour. Now prepare a paste of corn flour and butter or cream and add it into the concentrate. While adding flour or cream mixture, care should be taken to prevent coagulation of starch. Cooking on gentle flames is continued until desired consistency of the soup is obtained. Sugar and salt may be added now while removing the spice bag from the mixture (squeeze the spice bag tightly before removing it). Hot soup is packed in glass bottles or cans and sealed before sterilizing the bottles or the cans at 115 degree Celsius for 40-45 minutes. Sterilized bottles are then cooled and stored at ambient temperature in a dry place.

Figure 41: Tomato Soup

Nutrition in Tomato Soup: According to USDA, a detailed account of nutrition present in **canned, condensed tomato soup** is as given below:

Nutrient	Unit	Value/100 g
Water	g	80.79
Energy	kcal	66
Protein	g	1.46
Total lipid (fat)	g	0.44
Carbohydrate	g	15.22
Fiber	g	1.1
Sugars	g	8.23
Calcium, Ca	mg	13
Iron, Fe	mg	0.59
Magnesium, Mg	mg	14
Phosphorus, P	mg	31
Potassium, K	mg	562
Sodium, Na	mg	377
Zinc, Zn	mg	0.18
Vitamin C	mg	12.9
Thiamin	mg	0.042
Riboflavin	mg	0.015
Niacin	mg	0.858
Vitamin B-6	mg	0.086
Folate, DFE	µg	0
Vitamin B-12	µg	0
Vitamin A	IU	392
Vitamin E	mg	0.34
Vitamin K	µg	3.2

Tomato Puree and Paste: Tomato puree is the concentrated, tomato fruit pulp without skin or seeds and without added salt, and containing not less than 9% of salt-free tomato solids. Tomato fruit pulp is extracted from ripe tomatoes by using a pulper. Tomato pulp is then concentrated by cooking in an open cooker or in a vacuum pan to make tomato puree.

Tomato paste is different from tomato puree and it contains not less than 25 % tomato solids. Both puree and paste are prepared from fresh tomato juice and pulp which is extracted either manually using a sieve or by using a pulper. Tomato pulp is then strained to remove all particles before cooking it to desired consistency. Normally, an open cooker or a vacuum pan is used for cooking. A hand refractometer may be used to judge the required percent of tomato solids to be present in the end product. The end product or the final product is heated at 82-88 degree Celsius for a minute for homogenization and then this hot tomato puree or tomato paste is filled into cans and sealed. Sealed cans are then sterilized by immersing them in boiling water for 20-30 minutes. Sterilized cans are then cooled and stored at ambient temperature in a dry place.

Tomato Sauce and Ketchup: Ripe tomatoes are cleaned and crushed to extract juice. Juice is then sieved to remove all seeds and other materials to get a clear juice. Then, spices, salt, sugar and vinegar are added to the juice, with or without onion and garlic depending upon the consumer preferences. Good quality, tomato sauce should contain not less than 12% tomato solids and 25% TSS (total soluble solids).

Figure 42: Tomato Sauce

Preparation of Tomato Sauce: First step is to prepare a spice bag.

Ingredients for preparing a spice bag for 1 Kg tomato pulp: 50 grams of chopped onion, 10 grams of chopped ginger, 5 grams of chopped garlic, 5 grams of red chilli powder, 10 grams of cinnamon powder, 10 grams of large cardamom powder, 10 grams of aniseed powder, 10 grams of cumin powder, 10 grams of black pepper powder, and clove 5 numbers

For the process: Take 75 grams of sugar, 10 grams of salt, 25 millilitres of vinegar and sodium benzoate @0.25 grams/Kg of tomato pulp

Fully ripe, red tomatoes are harvested, washed and dried before using them for juice preparation. Clean tomatoes are then sorted based on their quality. Only good quality and juicy tomatoes are selected and trimmed before slicing them into 5-6 pieces. Sliced tomatoes are taken in a big jar and heated at 70-90 degree Celsius for 3-5 minutes to facilitate pulping process. Next step is the pulping process. Pulping or extraction of juice may be done either manually by using a stainless steel sieve (recommended for small quantities) or by using a pulping machine (pulper) for commercial-scale tomato juice production. Tomato pulp or juice is then strained to remove all particles. Pulp is then cooked with one third quantity of sugar. While cooking pulp, spice bag is dipped into it every now and then to allow the spices to mix with the cooked pulp. Pulp should be cooked until it reaches to one-third of the original quantity. Now remove the spice bag after squeezing its entire content into the pulp. Finally, add remaining quantity of sugar and salt. Cook this mixture for some more time.

Now it's time for the quality testing. A hand refractometer may be used to judge the required percent of tomato solids to be present in the end product. When the product is ready, add vinegar and sodium benzoate, a food preservative. Heat the final product at 82-88 degree Celsius for a minute for homogenization and then hot tomato ketchup or sauce is filled into bottles. Bottles are sealed using crown corking. Filled bottles are then pasteurized at 85-90 degree Celsius for 30 minutes. Bottles are then cooled and stored at ambient temperature in a dry place.

Nutrition in Tomato Sauce: According to USDA, a detailed account of nutrition present in **canned tomato sauce** is as given below:

Nutrient	Unit	Value/100 g
Water	g	91.28
Energy	kcal	24
Protein	g	1.2
Total lipid (fat)	g	0.3
Carbohydrate	g	5.31
Fiber	g	1.5
Sugars	g	3.56
Calcium, Ca	mg	14
Iron, Fe	mg	0.96
Magnesium, Mg	mg	15
Phosphorus, P	mg	27
Potassium, K	mg	297
Sodium, Na	mg	474
Zinc, Zn	mg	0.22
Vitamin C	mg	7
Thiamin	mg	0.024
Riboflavin	mg	0.065
Niacin	mg	0.991
Vitamin B-6	mg	0.098
Folate	µg	9
Vitamin B-12	µg	0
Vitamin A	IU	435
Vitamin E	mg	1.44
Vitamin K	µg	2.8

Canned Tomatoes: The process of sealing processed foods hermetically in containers after sterilizing them by heat for long-term usage is known as canning. Canning is one of the most popular food preservation techniques used today. Canning of fruits and vegetables and other highly perishable food items help reduce food waste.

Tomatoes meant for canning should be of good quality and they should be ripe and tender. The produce should be clean, free from foreign matter and dirt and devoid of any deformities and pest-disease infestations. The produce should be free from blemishers, cuts, or mechanical injuries. After choosing the right produce, they should be washed. Washing is done to clean the produce thoroughly. After washing, the produce should be air-dried.

Next step is blanching. Blanching is the most critical process in the entire canning process. Blanching is a brief heat treatment given to the vegetables before canning them. Major purposes of blanching are:

- Reduce microbial activity and contamination of products
- Remove undesirable elements from the products
- Remove undesirable taste and flavour from foods

After blanching, vegetables are dipped in cold water for keeping them in good condition. Cooled and air-dried, blanched products are then filled in containers or bottling cans. Before filling them in cans, these containers are sterilized. For can-filling, automatic, can-filling machines may be used. Hand filling is done for small-scale canning process.

Next step is brining. Brine is nothing but a concentrated solution of salt in water; brining is done to improve the flavour of canned vegetables and to serve as a heat-transfer medium. Common salt that is iron-free is used for brining purposes.

After brining '*exhausting*' is done. The process of removal of air from cans is known as exhausting. Exhausting is important because, corrosion of the tin plate during storage is avoided due to the removal of air. It minimizes discoloration of canned products by preventing oxidation process. It helps in better retention of vitamins and other nutrients present in canned foods. It prevents bulging of cans when stored in a hot climate at high altitudes.

Finally, sealing of cans is done. After filling cans, these containers must be sealed to make them air-tight. A can sealer may be used for sealing cans. In case of containers such as glass jars, a rubber ring should be placed between the mouth of the jar and the lid to make them air-tight. After sealing cans, they need to be cooled. Cooling may be done by any of the following methods:

- Dipping in Water: Dipping or immersing the hot cans in tanks containing cold water
- Flow of Cold Water: Letting cold water into the pressure cooker
- Spraying Cold Water: Spraying cans with jets of cool water
- Exposure to Cold Air: Exposing the cans to cool air

Storage of cans: Canned foods may be stored in wooden containers or CFB (corrugated fiber board) cartons. These foods should be stored in a cool, dry, hygienic place. Do not place cans at high temperature as it shortens the shelf-life of the products and also leads to the formation of hydrogen swell.

Figure 43: Canned Tomatoes

Tomato Marmalade: Marmalade is another fruit preserve made from sweet juicy tomatoes. A detailed account of cooking process of tomato marmalade is given below:

Ingredients Required:
1. 3 kg of ripe red juicy tomatoes: blanched in hot water for 5 minutes and immediately cooled in cold water; then skins removed
2. Juice and rind of 6 large lemons
3. 4 kg of brown sugar
4. 2 tablespoons of ginger juice
5. Water as per requirements
6. Salt to taste

Take some water in a large pan and place it on a low flame. Mix tomatoes, lemon rind, lemon juice and ginger juice in water and add salt as per taste. Cover the pan and let the mixture simmer until tomatoes are cooked properly. Add sugar and mix well with the mixture. Boil the mixture for about 15 minutes to bring the consistency of the mixture to the level of marmalade. When the mixture is properly set, remove it from the fire and cool it in a dry hygienic place. Pack the marmalade in an air tight sterilized glass jar for further uses.

Tomato Cocktail: Tomato juice is used for making various types of drinks and cocktails also. A simple recipe for making a tomato cocktail is given below:

First step is to prepare a spice bag. Ingredients required for preparing a spice bag for 5 liters of tomato juice: 0.25 grams of red chilli powder, 1.5 grams of cinnamon powder, 1.5 grams of large cardamom powder, 1.5 grams of cumin powder, 1.5 grams of black pepper powder, 1.5 grams of coriander seeds and clove 5 numbers

Other Ingredients: 60 millilitres of lime juice, 300 millilitres of vinegar (10 % acetic acid) and 45 grams of salt

Cooking Process: Strained tomato juice is cooked gently in a covered vessel for 20 minutes and spice bag is dipped into the juice all the while. After cooking, lime juice, salt and vinegar are added into it. Heat this mixture at 82-88 degree Celsius for a minute for homogenization and then hot tomato cocktail is filled into bottles. Bottles are sealed using crown corking. Filled bottles are then sterilized at 100 degree Celsius for 30 minutes. Bottles are then cooled and stored at ambient temperature in a dry place.

Tomato Chutneys: Tomato chutneys are easy to prepare and are very delicious. Some popular recipes for making different types of tomato chutneys are given below:

Spicy Tomato Chutney: First step is to prepare a spice mixture

Ingredients for preparing a spice mixture for 1 Kg tomato: 100 grams of chopped onion, 10 grams of chopped ginger, 5 grams of chopped garlic, 10 grams of red chilli powder, 10 grams of cinnamon powder, 10 grams of large cardamom powder, 10 grams of aniseed powder, 10 grams of cumin powder, and 10 grams of black pepper powder

Other Major Ingredients: 500 grams of sugar, 25 grams of salt, 100 millilitres of vinegar and sodium benzoate @0.5 grams/Kg of final product

Cooking Process: Fully ripe, red tomatoes are harvested, washed and sorted before blanching them for 2 minutes. Blanched tomatoes are immediately immersed in cool water to crack the skin which will in turn facilitate peeling process. Peeled tomatoes are crushed and spice mixture is added into it along with sugar. This mixture is cooked gently to desired consistency.

Now salt and vinegar is added into the mixture and again cooked gently for 5 minutes before adding sodium benzoate, a food preservative into it. This hot chutney is now filled into glass bottles and sealed before storing them at ambient temperature.

Red Tomato Chutney with Ripe Tomatoes

Take 1 kg of blanched tomatoes (sliced thinly) along with 100 grams each of red chilli powder and finely chopped ginger, garlic and onion. Take 250 ml of malt vinegar or apple cider vinegar, 500 grams of sugar, and 1 teaspoon each of green cardamom powder and clove powder (instead of clove powder, 4-5 whole cloves may be used).

Boil one small cup of water in a pan and add blanched tomatoes, garlic, ginger, onions and chilli powder and cook the mixture until it becomes soft and tender. While cooking, stir the mixture frequently and when the mixture is thick, add vinegar, and sugar along with cardamom powder and clove powder or cloves. Cook it again for 10 -15 under low fire. Finally add salt to taste before taking the chutney from fire and cooling it. Cool chutney may be stored in air-tight glass jars for several weeks and months.

Green Tomato Chutney – Sour

Take 4 green tomatoes (sliced thinly) along with 1 teaspoon each of mustard seeds, cumin seeds, and sesame seeds. Take 2 teaspoons of cooking oil, 2 tablespoon of dried coconut flakes and 6 green chillies (chopped finely), one pinch of asafoetida powder, and salt to taste.

Take a shallow frying pan and pour 1 teaspoon cooking oil. Heat the oil and fry the tomato slices along with sesame seeds, green chillies, and dried coconut. Grind the mixture together to a fine paste.

Now fry the mustard seeds in hot oil until they splutter, and then add cumin seeds and asafoetida powder. Pour this mixture over the tomato paste and mix well. Add salt to taste before consumption.

Sweet Green Tomato Chutney

Take 1 kg of ripe, green tomatoes (sliced thinly) along with 100 grams each of red chilli powder and finely chopped ginger, garlic and onion. Take 250 ml of malt vinegar or apple cider vinegar, 500 grams of sugar, and 1 teaspoon each of green cardamom powder and clove powder (instead of clove powder, 4-5 whole cloves may be used).

Boil one small cup of water in a pan and add sliced green tomatoes, garlic, ginger, onions and chilli powder and cook the mixture until it becomes soft and tender. While cooking, stir the mixture frequently and when the mixture is thick, add vinegar, and sugar along with cardamom powder and clove powder/cloves. Cook it for 10 -15 under low fire. Finally add salt to taste before taking the chutney from fire and cooling it. Cool chutney may be stored in air-tight glass jars for several weeks and months.

South Indian Tomato Chutney

Take 1 kg of ripe, red tomatoes (chopped finely) along with 100 grams of chopped onions, ginger and green chillies.

Now heat 1 teaspoon of cooking oil in a shallow pan and fry 1-2 whole dried red chillies; after that, add chopped onions, garlic and ginger and fry these ingredients until they turn slight-brown in colour. Now add chopped tomatoes and cook for a while until tomato pieces become soft. Take out this mixture from the fire and cool it before grinding the mixture into a fine paste.

Now heat another spoonful of cooking oil and crackle mustard seeds in the hot oil. Add 2-3 sprigs of curry leaves and 1-2 whole red chillies and fry it. Pour this mixture over tomato paste and add salt to taste. Mix well and serve it with south Indian food preparations such as *idli, dosa, upma,* and *vada.*

Tomato Chutney with Raisins

Ingredients Required:

- 2 kg of ripe red juicy tomatoes: blanched in hot water for 5 minutes and immediately cooled in cold water; then chopped finely
- 500 gm of brown sugar
- 500 gm of raisins: cleaned and chopped finely
- 500 ml of vinegar
- 100 gm each of finely chopped onion, ginger and garlic
- 50 gm of red chilli flakes
- Salt to taste

Cooking Process: Take a pan and place it on a low fire. Dissolve brown sugar in vinegar and pour this solution into the pan. Add chopped tomatoes, raisins, onions, ginger and garlic; add red chilli flakes; mix all the ingredients well. Add salt as per requirements.

Let the mixture simmer on a low flame; stir it occasionally. When the mixture is thick and becomes syrupy, remove it from fire. Cool the chutney in a dry hygienic place and pack it in an air tight sterilized glass jar.

Tomato Chutney with Date Fruits

Ingredients required:

- 2 kg of ripe red juicy tomatoes: blanched in hot water for 5 minutes and immediately cooled in cold water; then chopped finely
- 500 gm of brown sugar
- 500 gm of pitted date fruits: cleaned and chopped finely
- 500 ml of vinegar
- 100 gm each of finely chopped onion, ginger and garlic
- 50 gm of red chilli flakes
- Salt to taste

Cooking Process: Take a pan and place it on a low fire. Dissolve brown sugar in vinegar and pour this solution into the pan. Add chopped tomatoes, dates, onions, ginger and garlic; add red chilli flakes; mix all the ingredients well. Add salt as per requirements.

Let the mixture simmer on a low flame; stir it occasionally. When the mixture is thick and becomes syrupy, remove it from fire. Cool the chutney in a dry hygienic place and pack it in an air tight sterilized glass jar.

Tomato Pickles: Tomatoes are used for making different types of pickles too. Some popular recipes are given below:

Tomato and Onion Rings in Apple Cider Vinegar: Take large, fully ripe juicy red tomatoes. Wash tomatoes thoroughly and wipe them dry. Cut the tomatoes into thin round circular rings and keep it aside.

Now take large onions; peel them and wash them clean in running water. Wipe the onions dry before cutting them into thin circular rings. Now mix both tomato rings and onions rings together with a slight quantity of apple cider vinegar. Add salt to taste. Your instant tomato pickle is ready. If you like, you can add chopped herbs also into the pickle.

You may mix tempered spices (particularly fried mustard seeds and curry leaves) also for making this pickle more delicious and flavourful.

Figure 44: Tomato in Vinegar

Green Tomato Pickles: Take one kilogram of small, fully ripe, green or half-red tomatoes. Wash tomatoes in running water; wipe them dry. Cut each tomato into four equal sections. Smear the tomato pieces with sufficient quantities of salt. Keep it aside for few hours until tomato pieces lose sufficient amounts of water from it. Drain off the water just before the pickling process.

Pickling Process: Take a shallow pan and place it on full flame. Add 1-2 table spoons of cooking oil and heat it. Add 10 grams each of mustard seeds and aniseeds; fry these condiments. Reduce the flame and add tomato pieces along with spice powders (10 grams of turmeric powder, 50 grams of red chilli powder and a pinch of asafoetida powder). Cook the tomato pieces on a low flame until it is cooked properly. Add sufficient quantities of apple cider vinegar just a few minutes before stopping cooking. Make sure that tomato pieces are fully immersed in vinegar solution. Cool this pickle before storing it in air-tight containers. Keep the pickle refrigerated for future use.

Figure 45: Green Tomato Pickle: An Appetizer

Agrihortico Best Sellers

1. Advanced Hydroponics Technologies
2. Bulbous Vegetables
3. Growing Lilies, Tuberoses and Amaryllis
4. Curry Leaf Plant
5. Chile Peppers
6. Growing Edible Mushrooms
7. Mushroom Farming 21 Rules for Success
8. Growing a Home Garden
9. Growing Kale Leaves, Brussels Sprouts and Celery
10. Jalapeno Peppers
11. 3 Nutritious Specialty Cucurbits
12. Jerusalem Artichokes: Production and Marketing
13. Moringa, the Drumstick Tree
14. Asparagus Spears
15. Ginger, Turmeric and Indian Arrowroot
16. Growing Herbs for Aromatherapy
17. 5 Popular Perennial Vegetables
18. Bell Peppers
19. Common Medicinal Plants
20. Parsley
21. Patchouli herb
22. 21 Culinary Herbs
23. Brassica Vegetables
24. Mushrooms and Seaweeds
25. Nightshade Vegetables
26. Roots as Vegetables

www.ingramcontent.com/pod-product-compliance
Lightning Source LLC
Chambersburg PA
CBHW040316220526
45473CB00009B/2462